# Alcohol in Space

# Alcohol in Space

*Past, Present and Future*

CHRIS CARBERRY
*Foreword by* ANDY WEIR

McFarland & Company, Inc., Publishers
*Jefferson, North Carolina*

ISBN (print) 978-1-4766-7924-2
ISBN (ebook) 978-1-4766-3792-1

LIBRARY OF CONGRESS AND BRITISH LIBRARY
CATALOGUING DATA ARE AVAILABLE

Front cover images from NASA/Shutterstock

Printed in the United States of America

*McFarland & Company, Inc., Publishers*
*Box 611, Jefferson, North Carolina 28640*
*www.mcfarlandpub.com*

For Heather, Maddy, and Lucie

# Acknowledgments

This book would not have been possible without the support of numerous talented and knowledgeable people. First, I would like to thank my wife, Heather Carberry, as well as Rick Zucker for their assistance in proofreading and editing this book. I would also like to thank my employer, Explore Mars, Inc., for providing me the freedom to undertake this project.

I also had the pleasure of conducting interviews and receiving valuable feedback from the following experts (listed in alphabetical order): Clayton Anderson, Neeki Ashari, Paul Bakken, Gregory Benford, David Blackmore, Penelope Boston, Charles Bourland, David Brin, Samuel Coniglio, John Connolly, Octave de Gaulle, Mike Dixon, Jamie Floyd, Mark Forsyth, Ben Foster, Jonathan Frakes, Richard Garriott, Kellie Gerardi, John Grunsfeld, Gary Hanning, Jason Held, Scott Kelly, Jeff Kluger, Christophe Lasseur, Pascal Lee, Bill Lumsden, Taber MacCallum, Jeffrey Manber, Saralyn Mark, Ricardo Marques, Shohei Matsumoto, Ryan McElveen, Chris McKay, Bill Nye, Greg Olsen, Ian O'Neill, James Parr, Richard J. Phillips, Robert Picardo, J.C. Reifenberg, Brian Shiro, Peter Smith, John Spencer, Kirsten Sterrett, Neil deGrasse Tyson, Julio Valdivia-Silva, Reed Walker, Wieger Wamelink, Andy Weir, Raymond Wheeler, and Pete Worden.

I would also like to thank Wade Holler as well as Ardbeg, Budweiser, Cotton and Reed Distillery, Ninkasi Brewery, Suntory, Saber Astronautics, 4 Pines Brewing Company, the Library Company of Philadelphia, Steve Lansdale of Heritage Auctions, David Frohman of Peachstate Historical Consulting, Bryan Versteeg of Spacehabs.com, Pascale Ehrenfreund of DLR, the International Potato Center in Peru, the Open Space Agency, Jennifer Kennedy of GHO Ventures LLC, Phil Broughton, Christopher Shaffer, Jennifer Goldsmith, and many others.

Lastly, I would like to thank Charlie Perdue and the staff of McFarland.

# Table of Contents

# Foreword
# by Andy Weir

This book is a bit of a double-whammy for me. I'm a huge fan of space and space travel, of course. But what most people don't know about me is that I'm also an amateur mixologist. I love making mixed drinks and I have a whole wet bar set up in my home. I make everything from complex drinks (like the Ramos Gin Fizz, which requires me to manually shake it for eight minutes), to incredibly simple ones (like the Dark and Stormy, which is just rum and ginger beer).

My drink of choice is a rum old-fashioned: 2 oz of dark rum, ⅛ oz of simple syrup, 2 dashes of Angostura bitters, 2 dashes of orange bitters, over ice in a lowball glass. Tastes great! And it's fun to order in bars because that recipe I just gave you is only my personal approach. Every bartender has their own old-fashioned recipe and they vary wildly. And almost every one of them will raise their eyebrow in interest when you ask for a *rum* old-fashioned. They're supposed to be made with whiskey.

Now, I know what you're thinking: "Oh hey, a writer who drinks a lot. That's original." The truth is, I don't drink much at all. Maybe two or three drinks a week. But I really like tasting fine liquor when I do drink, and I love learning all the ins and outs of how to mix an excellent cocktail. It's similar to my interest in space, really. I love writing about space, and imagining adventures there, but I wouldn't go myself. I'm a happy earthbound misfit. If someone offered me a free trip to space, I'd ask if I can sell the ticket.

So, when Chris approached me to write an introduction to a book about space *and* liquor, how could I refuse?

One of the more common questions I get asked by fans of *The Martian* is, "Could Mark Watney have made vodka from some of those potatoes? Just to have a good time once in a while?" The short answer is "Only if he was willing to starve to death for vodka."

It takes almost eight kilograms of potatoes to make a bottle of vodka (that's seventeen pounds in Freedom Units). That works out to be almost a week of meals for our poor stranded astronaut. So no, Mark Watney would not waste food like that—especially when he was near starvation.

We're a little spoiled here on Earth when it comes to alcohol, because we have so many natural resources and plant life to make it with. In the end, as you'll see in this book, all you really need is a bunch of sugar and some yeast. But once we venture into space, we will no longer have the luxury of being utterly surrounded by fermentable crops.

So what happens when we colonize the moon? Or Mars? People aren't going to just stop wanting alcohol. And it'll probably be too expensive to import.

Idyllic views of the future always seem to come with the hidden assumption that human nature will change. That somehow, the flaws of mankind will just melt away amongst the awesomeness of living among the stars. People will abandon mundane flaws like booze and drugs, and also everyone will be super-efficient like some kind of environmentalist's dream. But that's never been the case as we march forward, so I don't see why it would happen in the future.

People are still going to want to drink. And they won't care if alcohol is an inefficient use of resources. So there's no question that we will manufacture alcohol in space in the future. The only question is how.

People will ferment anything. Good lord, *anything*. Oh sure, you know all about potatoes, beets, sugar cane, grapes, etc. But did you know people have also made liquor out of milk, raw chicken, eels, dead mice, brake fluid, and even human poop?

Yes, that's right. Mark could have skipped a step in that vodka plan and turned some of that poop directly into booze. "Pooze," if you will.

So, from humanity's history of gloriously turning anything into alcohol, I can only assume people would happily ferment algae. Algae is special because its life cycle is so fast that it can just keep doubling until it fills the available tank and/or consumes all the available nutrients and light. So we may see a future where a space-based company effectively makes booze from sunlight. If that sounds impossible, remember that's more or less what we do on Earth, too.

Of course, by the time we're living comfortably in space, there will probably be faster, more efficient ways to make alcohol. Some synthetic process for manufacturing ethanol that doesn't require the slow process of fermentation coupled with advances in artificial flavor technology. You could be drinking what tastes like some of the best rum Earth has to offer when it was actually made from a chemical process that starts with charcoal and water and had 1mL of flavoring added to it.

But who knows what the future holds? Well, maybe Chris does. So kick back, grab your drink of choice (rum old-fashioned for me, please), and enjoy this book about two of my favorite subjects.

*Andy Weir is the author of the best-selling and critically acclaimed novel* The Martian *as well as other works of science fiction. He is a lifelong space nerd and a devoted hobbyist of such subjects as relativistic physics, orbital mechanics, and the history of manned spaceflight. He also mixes a mean cocktail.*

# Preface

The concept of drinking and manufacturing alcohol in space may sound like science fiction, but in fact it is a very real phenomenon and has a fascinating history. *Alcohol in Space* is the first book written on this topic and tells the stories of the numerous occurrences of alcohol consumption in space and current space-related drink projects. *Alcohol in Space* also examines the history of alcohol in human culture, its treatment in science fiction, as well as technologies and innovations that will be required for manufacturing alcohol in space—specifically space agriculture.

Although the intent of this book is to provide a top-level overview of the various space drinks-related projects that have occurred or are currently underway, it is also based on extensive research that includes dozens of interviews with experts, accounts from nonfiction and fiction books, journal, magazine, and newspaper articles, blogs, videos, reports, television shows, and movies.

This book is both timely and important. As commercial space activities accelerate, this subject (both positive and negative aspects) will become more and more relevant as private individuals bring human customs, vices, and ceremonies into space. Since consuming alcohol is officially banned by most space agencies, there has been very little "official" research on this topic, including the physical and psychological impacts of alcoholic beverages in space.

Alcohol will almost certainly continue to accompany both private and government space travelers. As such, it is hoped that this book will initiate a more open discussion while, at the same time, telling a truly fascinating story.

# Introduction

*"After we find space aliens, my next question might just be, is there Burgundy wine elsewhere in the universe?"*
—Neil deGrasse Tyson[1]

Anyone who has joined friends or colleagues at a bar after work—or after a big event—can probably recall some really fascinating conversations taking place there. After everyone becomes more relaxed and has consumed a couple of drinks, inhibition sometimes slips away, and wild speculation prevails. The original spark for this book was inspired by a couple of decades of these types of barroom interactions with friends and colleagues in the space community. Subject matter would sometimes become a bit silly as we hypothesized "consequential" questions, such as, will we produce wine on Mars one day, and if so, what will it taste like? The home brewers among us would argue that beer would be a more realistic space drink, while the connoisseurs of spirits were convinced that distilled liquor was the obvious choice. After all, in *The Martian*, Mark Watney grew potatoes on Mars, which obviously means that vodka would be the drink of choice of brave Martian colonists in the future. Over the course of many years, these conversations would take some intriguing and sometimes ridiculous turns, but they were always relegated to the privacy of the pub.

Nevertheless, these gatherings often included astronauts, prominent experts in Mars science, and leaders in mission architecture design. While nobody outwardly proposed that space alcohol was anything but wild speculation, it was also clear that the discussions were fueled by genuine curiosity. After all, contemplating alcoholic beverages in space will probably have real merit "one day," but the time was not right to shift the topic to the mainstream—or at least the mainstream of the space exploration community. At the time, this was a completely valid determination. Plans for sending humans back to the Moon and on to Mars, as well as private efforts for space hotels and settlements, looked to all of us to be many decades away. In addition,

NASA and other space agencies prohibited the consumption of alcohol in space, thus, why would anyone bother elevating the topic of drinks in space beyond the confines of barroom brainstorming?

As it turns out, we were not the only bar patrons speculating on the prospect of off-world liquor. The same questions were being pondered by quite a few other people—and many of those people even moved these questions beyond mere pub banter. Alcohol producers and researchers are not waiting until settlements in space are a reality before they start planning the real future of space drinks. This book will tell the story of dozens of projects around the world—large and small—that have already engaged in a race to determine whether liquor can be produced in space, what impact space will have on the fermentation and aging process, and the safety and efficiency of alcohol production off planet Earth. Companies like Scottish whiskey maker Ardbeg as well as the Japanese whiskey company Suntory have conducted experiments on the International Space Station (ISS) to see how their products age in space. In 2017 American beer giant Budweiser announced plans to be the first beer manufacturer on Mars, and has already sent beer-related experiments to the ISS. Dozens of other similar experiments are taking place (or already have) around the globe—and more are materializing every year.

While these projects may be the basis for significant space-based alcohol production in the future, consumption of limited quantities of intoxicating beverages in space has been occurring for several decades. Even though astronauts and cosmonauts are prohibited from in-space imbibing, our space heroes of the past and present have, nevertheless, occasionally smuggled drinks on board their spacecraft. As will be shown in the pages of this volume, there are numerous stories of Russian cosmonauts carrying cognac and vodka to the former Mir space station and even the International Space Station, and the Russians are not alone. American astronauts have also carried—and consumed—adult beverages in space, including on the first mission to the surface of the Moon in 1969. While still on the lunar surface, Buzz Aldrin consumed wine as part of a communion ceremony—and Aldrin's lunar communion is far from the only case of Americans imbibing in space.

All these instances of drinking spirits in space have involved extremely small amounts of alcohol—only enough for an occasional toast. Any significant alcohol production in space will be almost entirely dependent on another capability, one that is also required for human life—agriculture. Any substantial quantities of indigenous alcohol will probably not be produced until agricultural efforts have sufficiently matured and reliable surpluses have been achieved. As the best-selling author of *The Martian*, Andy Weir, commented,

"Alcohol is just about the least-efficient use of crops. It takes a huge amount of otherwise edible food to make a small amount of alcohol. So, while it would certainly be a produced item in future space colonies, it wouldn't come about until they had a very plentiful food production capability."[2]

This will be vastly more challenging than historic exploration and settlement on Earth. When European settlers came to the Americas in the fifteenth century, they arrived at a continent teeming with life, one where crops could be plentiful without new techniques or technologies. In stark contrast, perfecting farming techniques on the Moon or Mars or even in low Earth orbit will require new capabilities and new ways of thinking about food production and other aspects of life. While most public attention gets focused on the development of big rockets and crew vehicles, this book will tell the fascinating story of the numerous space agriculture projects currently underway worldwide.

These amazing technological advances are vital for the future of adult beverages in space. However, it is also important to look at human culture and literature. This book will examine the long history and role of alcohol in human culture. As mentioned earlier, consumption of alcohol for pleasure and for ceremonial purposes has been practically synonymous with human culture since before recorded history—and has played a significant role in the exploration and settling of new frontiers. In short, it has been linked to the expansion of civilization and culture, a relationship that is not likely to end soon.

Works of fiction can also provide insight into this relationship and what the future may bring.

Some would argue that the works of science fiction can be just as accurate at predicting the future as the speculations of space agencies and scientists. Even if this notion is only partially correct, they would tend to indicate that space liquor is as inevitable in space as politics, death, and sex. Drinks and saloons in the stories of science fiction are not rare and often do not seem even remotely far-fetched. In film and television, dramas such as *Star Trek, Star Wars, Babylon 5,* and *Passengers,* as well as countless published works, seamlessly integrate various exotic drinks and cosmic watering holes into their storylines. Perhaps no other film scene better depicts more *literally* the phrase "alcohol in space" than the opening scene of the 1994 film, *Star Trek Generations,* which depicts a bottle of Dom Pérignon, vintage year 2263, tumbling through the vacuum of space and smashing on the hull of the USS *Enterprise* (NCC-1701-B), to christen that new starship. While we have no true idea what spaceships of the future will look like, this ceremony is a perfectly plausible, and even probable, scenario. It portrays human continuity and normalcy. It is a part of humanity that transcends time and cultural changes.

None of this will happen, however, without the development of critical infrastructure: the rockets, spacecraft, and space habitats that will enable this future. Visionaries, science fiction writers, and others have hypothesized about colonizing space for well over a century, but in the past such notions have been nothing more than science fiction and wishful thinking. Today, while true colonization of space is still probably years off, the development of such enabling technologies is well underway. On the government side of the equation, NASA and other space agencies have expressed a strong desire to mount international missions to the Moon and Mars. As such, for the past several years they have been working with industrial partners to develop the Space Launch System (SLS) launch vehicle (a heavy lift rocket) as well as the Orion crew capsule and other elements to enable human missions to deep space.

At the same time, so-called "new space" companies also have ambitious plans for the future of humanity. SpaceX founder Elon Musk has not been shy about his intent to build Mars colonies and claims that his company could get to Mars by the mid–2020s—and he is far from the only billionaire with major space aspirations. Amazon founder Jeff Bezos owns a rocket company called Blue Origin that has short-term goals of launching private astronauts into suborbital space, but hopes to send humans to the Moon and beyond within the next decade.

Hotels in space are the goal of Las Vegas hotel billionaire Robert Bigelow, who owns Bigelow Aerospace. Bigelow has been developing inflatable space station components for the last twenty years and has already launched two demonstration modules into orbit and provided an inflatable module for the International Space Station. Fellow billionaire Richard Branson's company Virgin Galactic plans to send paying customers as well as researchers into suborbit before 2021 and hopes to ferry customers to orbiting hotels and space stations and potentially begin point-to-point travel on Earth. These are just a sampling of hundreds of large, medium, and small efforts to make humanity more of a spacefaring race.

Alcohol production is far from their top priority, but former NASA Ames Research Center Director Pete Worden believes that consumption and manufacturing of alcohol in space is directly related to the progress of commercial space development, stating that "the more private sector activities there are in space, the less that these tight-laced government regulations are important. But large-scale production (of alcohol) and use won't happen until there [are] a lot of people living in space. I really expect it to take off when we see settlements on the Moon and Mars."[3]

The purpose of this book is not to proclaim that alcohol is always a positive factor or should be embraced by national space programs. As is the case with airline pilots or other professionals who are entrusted with the lives of many, limitations or restrictions are appropriate, but as we enter an age when private individuals start entering space, we would be in a state of denial (as well as irresponsible) to ignore the fact that alcohol will play a role in this new reality. For better or for worse, it would be prudent to better understand the implications of this fact. Unfortunately, as with the prospect of sex in space, drinking in space has become "taboo." Both are inevitable, but we find ourselves uncomfortable discussing the topic or conducting valuable research. For example, even though humans have consumed alcohol in space, no research has been conducted in space regarding the human body's ability to metabolize alcohol in that environment. Perhaps this is not as important as understanding the impact of bone loss in microgravity or how space radiation affects our physiology, but that does not mean that these areas of study should be ignored.

As such, if we are genuinely committed to making science fiction into a reality—of living and working in space—we can no longer restrict ourselves to the sterile notions of saintly space explorers who are free of vice. Space settlers will be real people. They will be extending human culture—and perhaps creating their own cultures. Human society can be messy. It can achieve wondrous accomplishments and good—but it can also stray into darkness. Alcohol in many ways represents this human dichotomy. As Iain Gately states in the opening of his volume, *Drink: A Cultural History of Alcohol,* "It is the most controversial part of our diet, simultaneously nourishing and intoxicating the human frame. Its equivocal influence over civilization can be equated to the popular character of Dr. Jekyll and Mr. Hyde. At times its philanthropic side has appeared to be ascendant, at others the psychopath has been at large."[4]

# 1

# A Brief History
# of Alcohol and Society

*"I have taken more out of alcohol than alcohol has taken out of me."*

—Winston Churchill[1]

Something magical happens in nature when glucose (sugar) and yeast interact. A biochemical reaction takes place—a reaction that has played a significant role in human civilization. The result of this transformation is ethanol and carbon dioxide. Ethanol is the basis for the intoxicating effects of the fermented and distilled beverages that humans have consumed for thousands of years. It is no exaggeration to say that the story of civilization is intimately linked with the story of these beverages. In fact, the roots of this relationship may have started millions of years prior to the beginning of "civilization." According to Dartmouth College biological anthropologist Nathaniel Dominy, "Our ape ancestors started eating fermented fruits on the forest floor, and that made all the difference.... We're preadapted for consuming alcohol."[2] This notion is also articulated by Reed Walker, the co-owner of Cotton and Reed Distillery and former NASA employee: "Humanity has always and will always drink booze. Food spontaneously ferments, and trace amounts of alcohol are found everywhere."[3]

As we begin to discuss the prospect of alcohol in space, it is essential to understand the long tradition of fermented beverages in human civilization and the many reasons why people consume these intoxicating drinks. To be clear, the following is not intended to be a comprehensive or complete history of alcohol and society. It offers, instead, a patchwork of historical anecdotes, providing some context to highlight the significant role alcohol has played in human culture.

Even a cursory glance at the ubiquitous nature of human culture and booze reveals that this relationship is highly unlikely to end when settlements are built on far-away worlds. "Drinks" have played a central role in the development of culture, social interaction, and have also had an undeniable role in religious expression. According to astrophysicist Ian O'Neill, "Saying 'Cheers!' while slamming our glasses together is a symbol of celebration, trust, and companionship."[4]

When and where the first human became tipsy from an intoxicating drink that had been intentionally fermented remains unclear, but the first *known* case of alcohol production took place roughly 10,000 years ago in the Neolithic era in northern China. A 2004 article in the *Proceedings of the National Academy of Sciences* noted that remnants of "mixed fermented beverages of rice, honey, and fruit"[5] were being produced in the village of Jiahu in the Henan province of China during that time. Little is known about how this early beverage was used within that society, but what is known is that fermented drinks incalculably influenced human events, including one of the key technologies that enable modern civilization.

According to one theory, the thirst for alcoholic beverages may have stimulated advances in early agriculture more than the need for food crops advanced that technology. According to Evolutionary Psychologist Robin Dunbar, the reasons that "humans started cultivating grains such as wheat and barley during the Neolithic [period] was not to make bread ... but to make gruel that could be fermented."[6] If true, that desire for a drink played a pivotal role in one of the core enablers of modern civilization. Without advanced agricultural techniques, civilization would not have become viable. This is further explained in a 2004 paper in the *Proceedings of the National Academy of Sciences* that states, "Because of their perceived pharmacological, nutritional, and sensory benefits, fermented beverages thus have played key roles in the development of human of horticulture, and food processing techniques...."[7]

Over time, virtually all sizable human cultures produced and consumed liquor for countless reasons that include ceremonial and religious purposes, medicinal uses, mind-altering effects, social and diplomatic activities, and pure pleasure. Exceptions to this rule have tended to be mostly cultures that were isolated. In Ian S. Hornsey's volume *Alcohol and Its Role in the Evolution of Human Society*, he noted that, "before the modern era, only the Eskimos, the peoples of Tierra del Fuego at the southern tip of South America and the Australian aboriginals apparently lived out their lives without the medical benefits or mind-altering effects of alcohol."[8]

In the case of the Inuits (Eskimos), climate was probably a major contributing factor. The Polar regions of North America tend not to be conducive to most of the traditional crops used in brewing and distillation. Otherwise, except for cultures where alcohol was forbidden for religious or other reasons, as will be shown, fermentation and consumption have generally been ubiquitous throughout human history, having as much of an impact as many important inventions and discoveries heralded through the generations.

## Ancient Civilizations

As civilizations began to spread across the globe, so too did the story of alcohol. In the Middle East, alcohol production appears to have started as early as 5000 BC in the Fertile Crescent between the Mediterranean and the Persian Gulf. The ancient Sumerians in Mesopotamia (now Iraq) were known to be prodigious brewers of beer, and had an enormous thirst for that beverage. As such, they had many sayings and proverbs related to beer, such as "He is fearful, like a man unacquainted with beer" and "Not to know beer is not normal."[9]

In his comprehensive examination of the history of alcohol and culture, *Drink*, Iain Gately describes many varieties of beer or "kash" that the Sumerian brewers produced. That culture actually identified over twenty varieties made from barley, wheat, and other grains. Beer had such stature in that society that Ninkasi became the goddess of the art of brewing.

Beer plays an important role in the well-known Sumerian poem, *The Epic of Gilgamesh*, which was written in roughly 2000 BC. In fact, it helps bring together two of the key characters, Gilgamesh and Enkidu. Enkidu is essentially living a feral existence away from civilization when he is lured out of the wild by a prominent harlot named Shamhat. Shamhat tells Enkidu, "Eat the food, Enkidu, it is the way one lives. Drink the beer, as it is the custom of our land."[10] Apparently, Enkidu instantly acquires a fondness for beer, consumes seven jugs of it, and "became expansive and sang with joy."[11] Shortly afterwards, he travels to the city of Uruk where he loses a wrestling match to Gilgamesh. However, this clash serves as a bonding moment between Gilgamesh and Enkidu and from that moment forward, they are inseparable companions. While this is just a story, it is but one of many examples of the importance that beer played in Sumerian culture.

### EGYPT

Both wine and beer were also an integral part of Egyptian life from the start of the Egyptian dynasties (approx. 3100 BC). Like in twenty-first-century America, beer, known as *hqt* in ancient Egypt, was considered a blue-collar drink that was consumed by the working class. Nonetheless, beer played a crucial role in Egyptian religious beliefs. According to Egyptian mythology, beer had once prevented the destruction of mankind. It should not come as a surprise to anyone that the sacred texts of many religions depict humanity as having a propensity for sinful behavior. Ancient Egypt was no exception. To punish the wickedness of humanity, Ra sent the lion-headed goddess Sekhmet to Earth to punish the sinners, attacking villages and ripping people apart and drinking their blood. Observing this carnage, Ra began to believe that Sekhmet's thirst for blood was literally "overkill," but Ra was unable to stop her rampage. Ra then devised an ingenious solution to this problem. He ordered 7,000 jars of beer to be colored red to resemble human blood and then had them strategically placed where he knew Sekhmet would find them. Sekhmet fell for Ra's trick and consumed all the blood-colored beer, becoming so intoxicated that she fell asleep for three days. When she awoke, she had become Hathor (another aspect of Sekhmet), the Goddess of motherhood, fertility, and the Milky Way—and no longer had the thirst for human blood. Through Ra's scheme, beer had literally saved humanity. This myth was celebrated annually at the Tekh Festival, also known as the Festival of Drunkenness.

Wine, or *irp*, did not have the epic backstory that beer had, but it did hold a higher place in Egyptian society. Like the perception of wine in modern society, it was a drink for the upper classes. Wine also had a patron god, Osiris, who was god of the dead, as well as life, rejuvenated vegetation, and wine. According to Iain Gately, this association between Osiris and wine grew in the later dynasties and in some rituals that were similar to later Christian rites. "His devotees, after prayers and rituals, would eat bread and drink wine in the belief that these were the transubstantiated flesh and blood of their divinity."[12]

The elevated status of wine in Egyptian society also caused special attention to be given to its production, resulting in the start of a categorization system that is reminiscent of modern methods for determining the quality and characteristics of wine. For example, the Egyptians categorized vintage years of wine, where it was produced, and who produced it. A jar of wine in the tomb of Tutankhamen had an inscription that described the wine as "Year 5. Wine of the House-of-Tutankhamun Ruler-of-the-Southern-On, l.p.h.[in] the Western River. By the chief vintner Khaa."[13] They also had a system for defining the quality of wines. Modern wine is often rated using a subjective numerical rating system, often appraising wines on a scale between 50 and 100. In contrast, the ancient Egyptians used a far more basic scale. A good wine would be called *nfr*. A very good wine would be called *nfr nfr* (or twice as good). And predictably, the next level of quality would be called *nfr nfr nfr* (or three times as good).

The Egyptians also appear to have understood the benefits of aging some wines. Additional evidence was discovered by Howard Carter in Tutankhamen's tomb in 1922. There were twenty-five vintages of wine found in this tomb and one of them appears to have been thirty-five years old at the time it was entombed with the young pharaoh. Indeed, wine was apparently a popular afterlife drink, as it was found in many Egyptian tombs. In fact, 700 jars of wine were found in a tomb in Abydos dating from the First Dynasty of Egypt.

While the Egyptians were not the first to consume wine, they played a major role in formalizing production, categorization, and practices that impacted winemaking for thousands of years

## GREECE

Egyptians may have raised the status and artistry of wine, but ancient Greece elevated the drink to a whole new level, and it became a powerful symbol of their society. Gately notes that "wine was omnipresent in Hellenic society. It was used as an offering to their deities; as a currency to buy rare and precious things from distant countries; and it was drunk formally, ritually, as a medicine, and to assuage thirst."[14] Gately also mentions that wine, or *oin,* also helped ancient Greeks differentiate themselves from other cultures who they deemed "barbario" or barbarians. Unlike wine enthusiasts of today, the Greeks did not drink their wine straight. Their wine would be watered down, often by at least two or three parts water to one part wine. Like many other rituals, this practice was based on an old myth. According to a physician named Philomides, many years earlier a band of Greeks were drinking wine along the shore when a violent storm broke out. The Greeks sought cover but had to leave their wine behind, exposed to the elements. When the storm ended, they found that the bowl of wine was now filled to the top with watered down wine. Not wanting to waste their drink, they tasted this storm-produced cocktail, and discovered that they liked it. Since it had been created during a thunderstorm, they attributed this heavenly mix to Zeus, who was both the king of the gods as well as the god of thunder.

That said, Zeus was not the final word for divine wine recommendations for the ancient Greeks. Wine was sacred enough to justify a patron member of Greek divinity. Dionysus was the god of winemaking and wine, grapes, fertility, and ritual madness. As with any self-respecting god, Dionysus allegedly inspired a following of fanatical "groupies." There was a cult of human female worshippers called Maenads who were quite dedicated in some unusual ways. In Mark Forsyth's book, *A Brief History of Drunkenness*, he describes the Maenads worshipping the wine god by "going out into the mountains wearing next to nothing and getting very, very drunk. Then they would dance and let their hair down and rip animals to pieces in a sort of terrifying Arcadian hen party."[15]

Based on the Greek fondness for alcoholic beverages, it should not be surprising that teetotalers were looked upon with suspicion in their society. Dionysus would often dissuade potential teetotalers from pursuing their destructive non-drinking existence. One example is a story of a teetotaling king who decides to outlaw "maenadism." In response, Dionysus makes his Maenads believe that the king was a lion. Since Maenads enjoyed dismembering animals as part of their worship rituals, they also proceeded to rip apart their king. Nobody wanted to have their bodies torn apart and their blood imbibed by a group of wild Dionysus fanatics, so this tale probably persuaded kings and commoners alike to drink alcohol.

There were numerous other rituals surrounding drinking in ancient Greece that did not involve dismembering animals. An important venue for drinking in Greek society was at a symposium. It should be noted, however, that a "symposium" had a different meaning in ancient Greece than the often-dull professional conferences that have usurped that word in modern Western society. To the Greeks, it was a drinking party. More precisely, it meant "drinking together."

A Greek symposium should not be mistaken as an antiquarian version of an American fraternity house, however. Symposia were often highly ritualistic and would be presided over by a "symposiarch" who directed the gathering. The symposiarch was elected, but was often the host of the party and was responsible for such tasks as choosing the wine and leading the various rituals. All symposia would start with a libation. Like the word "symposium," "libation" also had a different meaning in ancient Greece than it does in modern society; it was the practice of pouring wine on the ground to honor the gods, fallen heroes, and Zeus. The symposiarch would also decide upon the speed in which his guests would drink as well as the ratio of water to wine that guests would be served. According to Forsyth, "Nobody ever got drunk at a symposium by accident. At symposium you got deliberately, methodically and publicly drunk.... When the symposiarch said drink, you drank."[16]

Despite the apparent universality of alcohol in Greek (male) society, there were varying opinions on who should consume fermented beverages, how much should be consumed, and even what temperature it should be when served. Plato firmly believed in a minimum drinking age. He stated that "boys under eighteen shall not taste wine at all, for one should not conduct fire to fire. Wine in moderation may be tasted until one is thirty years old.... But when a man is entering his fortieth year ... he may summon the other gods and particularly Dionysus to join the old men's holy rite, and their mirth as well..."[17] It should also be noted that the only women present at these events were there for the pleasure of the male guests—such as dancers and courtesans. It would not be appropriate for Greek "ladies" to participate in such activities.

There were also opinions on how best to medically consume wine. The father of medicine, Hippocrates, believed that sustained consumption of warm wine caused "imbecility,"[18] whereas copious amounts of cold wine would cause "convulsions, rigid spasms, mortifications, and chilling horrors...."[19]

While Greece probably was not the first culture to ponder the positive and negative relationship between alcohol and human health, they unquestionably elevated this topic to a new level of scholarship. Many of the discussions that began in Greece are still relevant today.

## ROME: VENI, VIDI, VICI VINUM

The Romans borrowed many elements of Greek culture as they expanded their empire, including Roman versions of the Greek pantheon of gods. However, in the early years of Rome, drinking habits deviated considerably from Greece. In fact, the Romans endured several periods of near sobriety and would often brutally purge followers of Bacchus, the wine god (their version of Dionysus). This initial ostracism of drink was said to have started with the founding of Rome when the mythical ruler Romulus would use milk as offerings to the gods rather than wine, but this sobriety would not last. Eventually, as Rome expanded its empire, and as more and more foreign influences entered Roman life, they embraced a drinking culture similar to what Greece had enjoyed.

Like the Greeks, the Romans diluted their wine with water. In addition, they believed that only barbarians drank straight wine. Of course, adding wine to water would hinder the growth of bacteria and pathogens in the water and would make it safer to drink. The Romans would not have known the scientific justifications for this practice, but they would have noticed that fewer people got sick drinking heavily diluted wine rather than straight water.

Even as Roman drinking habits grew, there were a number of authors advocating for moderation, including the Roman poet Horace, who lived by the motto "Let Moderation Reign!"[20] That said, Horace was as much an admirer of wine as most of his contemporaries. In his *Epistles V,* he writes,

"It (wine) unlocks secrets, bids hopes to be fulfilled, thrusts the coward onto the battlefield, takes the load from anxious hearts."[21]

In 79 AD, however, the Roman wine industry was struck with a major and unexpected blow; not by the changing ideology of their government, but by the eruption of Mt. Vesuvius, famous for burying the city of Pompeii under meters of volcanic ash. In Hugh Johnson's *The Story of Wine*, the author writes that within the miles of destruction around the volcano, "Rome's principal source of wine went with it: the 78 vintage was destroyed, the 79 never made."[22] The excavation of Pompeii shed a lot of light on Roman drinking habits. Within Pompeii, the volcanic ash covered an estimated 200 bars in that city, including one single block that had eight bars. Clearly the more temperate Roman attitude toward drinking of previous centuries was truly a thing of the past; and so too was Pompeii until it was excavated many years later.

## Other Gods of Wine and Beer

Alcohol deities such as Dionysus, Ninkasi, Ra, and Osiris tend to get much of the modern-day accolades as the immortals of drink, but most major cultures had gods devoted to wine and/or beer. In Norse mythology, the god Aegir is intimately connected with beer. While he was the god of the sea, he was also known for wild drinking parties where Odin, Loki, and Thor would consume enormous amounts of beer.

Alcohol ("jiu" in Mandarin) has a longer known history in China than anywhere else on Earth and has played an important role in their culture, so much so that it is called the "water of history." As such, the Chinese also have had deities of the "drink." There is a Chinese legend that, over 4,000 years ago, the wife (sometimes referred to as his daughter, sometimes as a servant) of an emperor in the first dynasty not only invented alcohol but later was elevated to divine status. Yi Di apparently concocted a variety of rice brews and presented it to the emperor. According to some accounts, the emperor loved it, but did not think anyone else could handle it—thus he banned the beverage and exiled Yi Di. However, Yi Di landed on her feet (or possibly wings) since she subsequently became a goddess.

The Hittites, an ancient culture that lived in what is now Turkey (from roughly 1600–1200 BCE), had a wine god named Teshub. In many ways, Teshub possessed an even higher level in the Hittite pantheon than the wine gods in other cultures. Teshub was the son of the king of gods, Kumarba, who eventually overthrew his father to himself become king of the gods. Having the king of the gods as a patron deity probably provided a lot of perks for Hittite winemakers.

Just a few of the other countless gods who reigned over alcoholic beverages include the Aztec god Tezcatzontecatl, who was the god of fertility, drunkenness, and pulque (Pulque is a fermented beverage made from the sap of a maguey plan—related to the Agave plant). The Zulu god Mbaba Mwana Waresa reigned over beer as well as rain, agriculture, and fertility, and the Celtic god Sucellus held dominion over forests, agriculture, and beer.

Few modern cultures pray to gods of beer, wine, and spirits, but the universality of deities of drink show the significant role and status that alcohol played in ancient cultures around the world.

## Modern Religion and Demon Alcohol

The influence of "demon alcohol" was not limited to ancient religion. It has played—and continues to play—a significant role in the traditions of many modern faiths. At times this relationship has been fraught with conflict and even contradictory opinions regarding what the role of liquor should be. Christianity, Judaism, and Islam are examples of modern religions with a significant amount of shared theological heritage, but all three of these faiths have contended with the role of alcohol in different ways.

### CHRISTIANITY

As seen in the story of the last supper of Jesus Christ, wine is at the core of one of Christianity's most sacred ceremonies. Communion (the Eucharist) emulates the Last Supper of Jesus Christ, in which bread represents the body of Christ and wine represents his blood. The Last Supper story is far from the only reference to alcohol, particularly wine, in the Bible. One can argue that the second most prominent story regarding alcohol in the New Testament occurs in John 2:10 during which Jesus famously turns jars of water into wine at a wedding at Cana in Galilee. Apparently, Jesus was not content to make a poor-quality wine. Upon tasting the wine that Jesus had made, the governor of the feast stated to the bridegroom that "every man at the beginning doth set forth good wine; and when men have well drunk, then that which is worse: *but* thou hast kept the good wine until now."

Beyond the ceremonial uses of alcohol, various Christian religious orders have produced significant amounts of liquor. Monks became famous for brewing beer, wine and champagne, and spirits. However, this tradition has also resulted in often non-flattering stereotypes of drunken monks who seem more devoted to drink than to faith. Nonetheless, they played a significant role in the production, advancement, and distribution of alcoholic beverages around the world. Since wine was at the heart of the most sacred ceremonies in the Christian faith, the growth and expansion of wine production went hand in hand with the expansion of the faith. In the early 200s AD, St. Clement of Alexandria outlined the etiquette that should be followed by Christians regarding drinking in a work called *Pedagogia*. Like many commentators on drinking throughout the ages, he advocated moderation. He is also clear that there is an obligation by Christians to drink wine because "to drink the blood of Jesus is to become partaker of the Lord's immortality."[23] Over a century later, St. Martin of Tours is reported to have spread winemaking throughout the Loire Valley in France.

However, it was not just wine that was spread by Christianity. Centuries later, the church was instrumental in expanding distilled liquor. This was the case in Ireland. While the Irish did not invent distillation, there is strong evidence to suggest that Irish monks were the first to create whiskey as we know it. The word "whiskey" is based on the Gaelic word "uisce beatha"—which means "water of life." It should also be noted that the Latin phrase *aqua vitae* also meant "water of life" and came to be associated (although not exclusively) with whiskey in this era. The first known written reference to whiskey appeared in 1404 in the *Annels of Clonmacnoise*, which documented the history of Ireland up until 1408. Later that century the first known reference to "scotch whiskey" appeared in the Exchequer Rolls of 1494, which showed that King James IV of Scotland had sent malt to Frier John Cor, "by order of the king, to make aquavitae."[24]

This strong (almost symbiotic) relationship between Christianity and alcohol became more complicated in later years. A taboo began to grow around the use of alcohol in some sectors of the faith. This was particularly the case in the United States, where Christian groups played a pivotal role in the temperance movement in the nineteenth century and today. To this day, many evangelical Christians shun alcohol. In fact, a 2011 Pew Research poll found that 52 percent of evangelicals polled thought that alcohol consumption is incompatible with their religious beliefs.

In another sphere of the Jesus Christ–based faiths, the Church of Latterday Saints (otherwise known as Mormons) have even more clearly defined rules regarding alcohol. According to the "Words of Wisdom" in *The Book of Mormon*, "5. That inasmuch as any man drinketh wine or strong drink among you, behold it is not good, neither meet in the sight of your Father, only in assembling yourselves together to offer up your sacraments before him." It goes on to say, "7. And, again, strong drinks are not for the belly, but for the washing of your bodies." In this same section of "Words of Wisdom," tobacco and "hot drinks," such as tea and coffee (as well as any caffeinated drink), are not permitted.

## JUDAISM

Judaism has not been as conflicted with regard to alcohol use as Christianity has been. Although Jews as a group tend not to overindulge in alcohol use in their personal lives, wine does play a key role in many Jewish religious holidays and ceremonies. This includes *kiddush*, a prayer said at the beginning of the Sabbath and during religious holidays, and which involves a blessing that is recited over a glass of wine, as well as during Havdalah, a ceremony marking the end of the Sabbath. Wine is also a key part of wedding ceremonies, the Passover Seder, and in circumcision ceremonies. In fact, during the circumcision ceremony, a drop or two of wine may be placed in the infant's mouth to serve as a type of sedative.

The Old Testament of the Bible contains numerous references to wine as being a sacred drink. Judges 9:13 states that wine "brings joy to God and man." There is even a blessing of the wine that is recited during many religious services that translates to "Praise to You, Adonai our God, Sovereign of the universe, Creator of the fruit of the vine." After spending over 300 days on a boat with innumerable pairs of animals, one can sympathize with Noah after landing on Mt. Ararat, when he "drank of the wine, and was drunken, and was uncovered in his tent."

This is certainly not to say that the Bible advocated unbridled imbibing. There are numerous warnings about alcohol. Proverbs 20:1 (in the King James Version) states that "wine is a mocker, strong drink is raging: and whosoever is deceived thereby is not wise." And in Proverbs 23:32, Solomon describes wine: "At the last, it biteth like a serpent and stingeth like an adder."

Nonetheless, Judaism has not been as prone to moralizing over alcohol and in fact has also played a major role in advancing the wine/alcohol trade. According to Iain Gately, while Jews have not been known for immoderate drinking, he states that they were "a wine-drinking culture, indeed, had been enamored of the grape for millennia prior to the foundation of Rome. Archeological evidence suggests that their Semitic predecessors had carried out an extensive wine trade with Egypt...."[25]

This connection to the alcohol trade did not end in Rome. It was one of the few trades that were open to Jews, and Jewish merchants were engaged in it for the centuries that followed. One reason for this was directly interlinked with requirements of their faith. In her book *Jews and Booze*, Marni Davis explains that "Jews are linked to alcohol production and consumption by the dietary regulations of kashrut, which requires Jews to use wine in their religious rituals and forbid consumption of wine produced or handled by non–Jews."[26]

These religious requirements not only created a market within the Jewish community, but also put Jewish traders and producers of wine and other alcoholic beverages in a perfect position to cater to non–Jewish consumers. As will be discussed later in this chapter, this relationship would become very complicated in advance of and during Prohibition in the United States.

## ISLAM

It is common knowledge that alcoholic consumption is either *haram* ("forbidden") or greatly frowned upon in many modern Islamic countries. In fact, the Islamic holy book, the Quran, contains several references to alcohol and consumption of that type of beverage. The Quran does not universally condemn alcohol throughout its pages, however. According to an article in *The Washington Post* by Khaled Diab, "From Islam's very inception, there has been a debate about what exactly the Koran's passages on drinking prohibit. Although the majority opinion holds that the intoxicant—alcohol itself—is banned, a minority view is that it is intoxication—getting drunk—that is forbidden."[27]

However, there are unquestionably strong words of caution in the Islamic holy book. In a translation (Shakir) of Quran 2:219 one passage reads, "They ask you about intoxicants and games of chance. Say: In both of them there is a great sin and means of profit for men, and their sin is greater than their profit." Stronger language appears in a translation (Sahih International) of Quran 5:90 which states, "O you who have believed, indeed, intoxicants, gambling, [sacrificing on] stone altars [to other than Allah], and divining arrows are but defilement from the work of Satan, so avoid it that you may be successful." Nonetheless, while wine should not be consumed during life, wine is held in such high esteem that it is reserved for the afterlife.

Despite the warnings and laws that have been adopted, alcohol can be found in many Islamic countries—particularly countries with a healthy tourist trade, such as Egypt. Egypt has tended to be more flexible with regard to alcohol consumption. In the nineteenth century, British scholar Edward William Lane observed that many Muslims in Egypt "drank wine, brandy, [etc.] in secret; and some thinking it no sin to indulge thus in moderation, scruple not to do so openly."[28] Egypt also has had a tradition of brewing beer—and other beverages—for thousands of years. The biggest producer of beer in Egypt is Al-Ahram, which has been brewing since 1897. While it was founded by a Belgian company—and has shifted ownership through the years (it is now part of the international Heineken Group)—it runs five plants in Egypt and employs over 1,800 people in that country.

In other countries, such as the United Arab Emirates (UAE)—particularly Dubai—tourists have plenty of options for having a drink. Tourist hotels in Dubai are well known for their numerous bars. Nonetheless, alcohol tends to be tightly controlled and in most cases reserved for outsiders (non-believers) who are visiting these countries or are contracted to work in these regions. Despite the accessibility of alcohol to visitors, foreign imbibers still need to be careful and not overindulge. Westerners have been arrested for drinking in public or violating permissible drinking rules in Dubai and other parts of the UAE.

## Shintoism

Eastern religions also have a fondness for inebriating beverages. Sake—or rice wine—is considered the drink of the gods in the Shinto religion and is believed to have first been brewed in roughly 300 BC. As such, it plays a significant role in many religious ceremonies. According to Tetsuo Hasuo from the Japan Sake Brewers Association, "In Japan, sake has always been a way of bringing our gods and people together."[29] In fact, in many of Japan's older texts the word used for sake is "miki" and is written with the characters used for both "god" and "wine."

Sake is an indispensable part of various festivals (matsuri) that take place in Shinto shrines where they offer food and drinks to the gods. In a 2000 speech called "Sake: Drink of the Gods—Drink of the People," Japanese sake brewer Daimon Yasutaka explained, "Naturally, the god does not actually eat and drink these, but they are offered with a spirit of modesty. In this way, these offerings of food and drink are infused with the blessing of that god, and filled with grace. Later, the people partake of this food and drink and are themselves filled with the blessings and grace of god."[30] Yasutaka also notes that "the most important of all the food and drink items offered to a god is sake. Sake itself is a blessing from the gods, and is created by brewing rice, another gift from the gods. Not only that, the light inebriation we feel when we drink sake is a special feeling that can be likened to being transported to another world, so it is viewed as a very special drink amongst the offerings."[31]

Unlike some Western religions, Shintoism is not as conflicted about its embrace of alcohol within its ceremonies and is not shy about this fact. Sake barrels ("kazaridaru" or decoration barrels) are often stacked outside Shinto shrines. Even for the more liberal churches in Christianity, it would be considered highly inappropriate to stack beer, wine, or whiskey barrels outside their places of worship in this way.

## Exploration and Settlement

Some of the most significant phases in the story of humanity's relationship with alcohol have occurred during the great voyages of exploration and settlement to the far reaches of the Earth. However, inebriation only represents a part of this story. In fact, beer, wine, and spirits were required for crew/explorer health. Water supplies were often suspect, particularly on long ocean voyages when fresh and clean water were hard to come by. As such, carrying significant supplies of booze for the passengers and crew was a requirement; it was quite literally for their health. One of the founders of modern military medicine in the eighteenth century, Sir John Pringle, commented on this, stating, "It hath been a constant observation, that in long cruizes or distant voyages, the scurvy is never seen whilst the small-beer holds out, at a full allowance; but that when it is all expended, that ailment soon appears. It were therefore to be wished, that this most wholesome beverage could be renewed at sea; but our ships afford not sufficient convenience."[32]

A notable example occurred during the first circumnavigation of the world. When Ferdinand Magellan set out on his expedition, he was extremely well stocked in alcoholic beverages. According to Iain Gately, "Magellan spent more on sherry than on armaments; indeed, wine rations cost nearly twice as much as his flag ship, the *San Antonio*,"[33] but this would have been quite logical. Outfitting a voyage to circumnavigate the globe would have been a monumental challenge. Magellan's expedition would certainly need to be well armed for their trip (as illustrated by the fact that Magellan died after being struck by a poison arrow during a skirmish in the Philippines), but armaments would rarely be used. Alcoholic beverages, however, would be consumed daily by the crew and therefore would need to be plentifully stocked. Given the length and hardships of such a voyage, the drinks would not only have been essential to maintain crew health, but also to maintain their social and psychological well-being. Indeed, the psychological benefits of alcohol on these voyages must not be ignored. Alcohol could often provide a simple remedy to the stress of long and dangerous ocean voyages.

Numerous modern studies have shown that the early explorers and sailors may have had the correct idea. For example, a 1985 study by clinical psychologist Cynthia Baum-Baicker outlines several benefits of moderate alcohol consumption—benefits that would be highly relevant to crew members on pre-industrial revolution voyages. Some of these benefits include stress reduction, mood enhancement, cognitive performance, and reduced clinical symptoms, primarily of depression. All of these factors would have clearly benefited individuals who were living in harsh conditions and often isolated from society and their families—for months or even years at a time. More recently, Stanton Peele and Archie Brodsky outlined that "positive drinking" was shown to have benefits such as "mood enhancement, stress reduction, sociability, social integration, mental health, [and] long-term cognitive functioning."[34] This is not to imply that there were not numerous negative aspects to alcohol consumption and exploration, but too often consumption of booze in these types of circumstances is dismissed as an entirely negative practice.

While abuse clearly took place (particularly by today's standards), it is entirely possible—based on these studies—that the psychological and social benefits of having some alcohol each day may have saved many of these expeditions from internal collapse.

## FRANCE, GERMANY, AND RUSSIA

Few people are better known for their appreciation of fermented drinks (particularly wine) than the French. Indeed, winemaking in France is estimated to go back more than 2,000 years. Some argue that the hillsides of Hermitage in the Rhone Valley were "the first with vines around 600 B.C.—Syrah vines, that had come up the river from Marseille with the first band of invading Phocaean Greeks."[35] Whether this is true or not, wine and other forms of alcohol became (and remain) a core part of French culture.

Beyond the well-known stereotype of the French consuming wine and cheese at sidewalk cafes, wine is genuinely hardwired into French culture and has impacted their history in many ways. In fact, alcohol played a role in the outbreak of the French Revolution in 1789. At the time, the average French citizen, whether he/she was a member of the aristocracy or of the lower classes, drank a substantial amount of wine compared with modern standards. Some estimates put the average annual per person consumption of wine in Paris at 300 liters. As such, taxes on these beverages were a major source of income for the crown. According to historian Noelle Plack, "More revenue came from taxes on alcoholic beverages entering Paris than from all other commodities combined. The rate of tax was also very high, with the price of a barrel of wine tripling as it passed through the city's tollgates. These taxes and the customs barriers of Paris became the focus of revolt in July 1789."[36] In the days leading up to the storming of the Bastille on July 14 of that year, 40 out of 54 tollgates around Paris were destroyed by angry crowds.

The French Revolution was not the only anti-monarchical uprising influenced by booze. Nobody would argue that alcohol (or the lack thereof) was a primary cause of the Russian Revolution, but it also (probably) played a role in the downfall of the Russian monarchy. In 1914, Czar Nicholas II banned alcohol in Russia as that country mobilized for World War I. Since one-third of government revenue came from taxation on vodka, this was a curious strategy by the Czar, but in a document sent to the Russian Minister of Finance, Nicholas II explained that government income should come "not from the sale of something that destroys 'the spiritual and economic powers' of the people but from other, healthier sources."[37]

Nevertheless, while per capita alcohol consumption was substantially less than that of France and some other countries, the Russians were still quite attached to vodka. Like the experience in the United States a decade later, bootlegging and other illegal methods of obtaining alcohol became quite common. While there were some benefits to this prohibition—the crime rate went down and general health appears to have improved—there also were numerous riots, including the destruction of 230 former saloons in 1914 by people demanding vodka.

How much this actually influenced the downfall of the Russian monarchy is questionable, but three years after Nicholas declared Prohibition he was ousted from his throne when the Russian Revolution(s) took place. Anyone who believed that the Bolsheviks would repeal the Prohibition decree would be greatly disappointed. Vladimir Lenin opted to keep Prohibition in place. It was not until after Lenin's death, when Joseph Stalin took the reins of the Soviet Union, that Prohibition was repealed in Russia (now the heart of the Soviet Union).

Booze did not play a role in a bloody revolution in Germany, but it certainly has played a substantial role in German culture. Curiously, one of the greatest symbols of the German love of beer began with a wedding and a horse race. In 1810, King Maximilian I Joseph of Bavaria announced a mid–October festival in Munich to celebrate the wedding of his son Prince Ludwig to Princess Therese of Saxony-Hildburghausen. The festival would offer free beer and food, dancing and music, an agricultural festival, and a horse race. Although a wedding had motivated the original event, it was so popular that they continued orchestrating October festivals. Over time, the festival was expanded—and the horse race was dropped. Eventually, beer became the central theme of this event that became known as Oktoberfest. Oktoberfest is now one of the premiere drinking events on Earth and one of the best known (if not *the* best known) celebration of German culture.

## EARLY UNITED STATES

As with most other expansions of civilization, booze also influenced the settlement of the British colonies in North America and may have even played an unexpected role in the creation of the Thanksgiving holiday in the United States. Virtually all ocean-going vessels at the time carried beer and other alcoholic beverages. As stated earlier, these supplies were essential for crew health and mental wellness. In fact, *The Mayflower* ordinarily transported wine and dry goods, but was hired—along with a ship called *The Speedwell*—to ferry the Pilgrims to North America. However, early in its voyage, *The Speedwell* began to take on water and had to turn back. As a result, all the Pilgrims had to squeeze themselves on *The Mayflower*.

After experiencing extremely rough seas as they approached America, they landed in Massachusetts rather than Virginia, and as winter approached, the passengers and crew faced a serious problem. *The Mayflower* was running low on a vital supply—beer. The reason this was such a problem was the fact that they did not believe it was safe to drink water on the ship or from the shore. As beer supplies diminished, so too did trust between the Pilgrims and the crew of *The Mayflower*. According to Gately, "The shortage of beer was a point of friction between them (the Pilgrims) and the crew of *The Mayflower*."[38] Eventually the situation deteriorated to the point where the crew would not provide beer to sick passengers. As recalled by William Bradford, "As this calamity fell … the passengers that were to be left here to plant … were hastened ashore and made to drink water that the seaman might have more beer, and one [Bradford] in his sickness desiring but a small can of beer, it was answered that if he was their own father he should have none."[39] How much the shortage of beer impacted the destination and overall fortunes of the Pilgrims is hard to determine, but it was certainly a major concern at the time. It should also be noted that while the first Thanksgiving meal lacked most of the traditional foods we associate with it today, it did have beer, wine, gin, and brandy on hand.

Clearly, in the case of "demon whiskey," it would be a misnomer to state, as many people do, that people who oppose alcohol consumption for religious purposes are "puritanical." In reality, being puritanical would be an invitation for a cold (or probably warm) brew.

Years later, the value of intoxicating beverages even appeared in the writings of some Puritan ministers. One of the most prominent ministers of his generation was Increase Mather. In his anti-drinking sermons, *Wo to Drunk-ards*, Mather is often quoted as saying that wine was "a good creature of God and to be received with thankfulness."[40] However, he cautioned that "the abuse of drink is from Satan, the wine is from God, but drunkard from the Devil."[41] Like most earlier references, alcohol was deemed as necessary, but was to be consumed in moderation. According to Bruce Bustard of the National Archives, "while drunkenness was seen as disruptive to community, social occasions such as weddings, barn raising, elections, christenings, and funerals were opportunities to indulge."[42]

Over a century later, when the influence of Puritanism had diminished, the dangers of excessive alcohol consumption came under greater scrutiny. In his 1785 pamphlet, *An Inquiry into the Effects of Ardent Spirits upon the Human Body and Mind*, Dr. Benjamin Rush outlined the dangers to health and prosperity. By "ardent spirits" he meant "liquors only which are obtained by distillation."[43] Rush was a highly respected physician who was a signor of the Declaration of Independence and was also close friends with many of the other signors, including John Adams and Thomas Jefferson.

As indicated above, Rush was only concerned with "ardent spirits" or distilled alcohol. In fact, he considered non-distilled alcohols such as beer and wine to be healthy parts of the diet. According to Rush, "They are, more-over, when taken in moderate quantity, generally innocent, and often have a friendly influence upon health and life."[44] He also recommended—in addition to "simple water"—wine, malt liquor, cider, and coffee as alternatives to intemperance. Wine in particular he describes as "cordial and nourishing. The peasants of France, who drink them in large quantities, are a sober and healthy body of people."[45] Essentially, he was stating that as long as you drank beer and wine, you could avoid alcoholism. Whereas ardent spirits could cause—among many other things—"profane swearing … certain immodest actions"[46] in otherwise "chaste and decent"[47] women, as well as in some cases "roaring, imitating the noises of brute animals, jumping, tearing off clothes, dancing naked,"[48] and many other negative actions. Although the more con-servative standards of Puritanism had lessened over the previous decades, tearing ones clothes off and dancing naked in public (or elsewhere) would still have been quite scandalous—as it would be today.

To help avoid such unfortunate occurrences, Rush even created a chart that he called "A Moral and Physical Thermometer" that outlines the "health and wealth" benefits of temperance—which included drinking alcohol in moderation, but also the precise consequences of intemperance. Consump-tion of ardent spirits at the lower reading on his thermometer would result

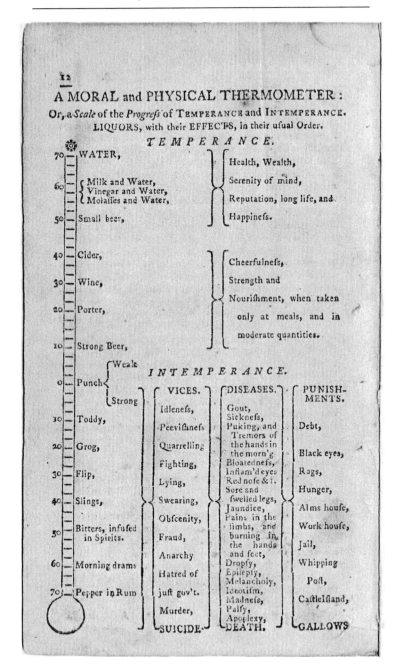

A "Moral and Physical Thermometer" designed by Benjamin Rush to illustrate the dangers of drinking distilled alcohol or "Ardent Spirits" (The Library Company of Philadelphia).

in debt or jail, but consumption on the higher end of this thermometer would lead to the gallows.

Dr. Rush was not the only Founding Father of the United States to have strong feelings about strong drink, but some of the others were actually purveyors of spirits. In fact, the first president of the United States was among the founding fathers of alcohol production in United States. George Washington ran one of the largest distilleries in the new world, producing brandy and whiskey. In 1799, this distillery produced 11,000 gallons of whiskey.

Thomas Jefferson is often described as the founding father of the American wine industry even though there is little evidence that he ever produced any wines of his own—at least of a quality that he would be willing to share. Jefferson was a great collector of wine and he also owned vineyards in Virginia. He believed that wine was a necessary part of healthy living. In a letter to John F. Oliveira Fernandes in December 1815, he wrote, "Disappointments in procuring supplies have at length left me without a drop of wine.... Wine from long habit has become an indispensable for my health...."[49] Like Rush, he was a critic of "ardent spirits." In an 1818 letter to Jean Guillaume Hyde de Neuville, he stated, "No nation is drunken where wine is cheap; and none sober, where the dearness of wine substitutes ardent spirits as the common beverage."[50]

Although Jefferson was not successful at winemaking, he still played a large role in the future of the Virginian and American wine industries when he began planting his vineyards. He started off on a (very) small scale at first. In 1771 and 1773 he noted planting unidentified grapes and vines at Monticello. As of 1778, Jefferson noted a vineyard that measured only 90 by 100 feet. Clearly, this was not the scale of winegrowing that one would consider the start of a winemaking empire.

Jefferson also gave a far more substantial amount of land in 1773 (193 acres) to Italian viticulturist Philip Mazzei who attempted to grow European vines near Monticello. Mazzei had been convinced to come to Virginia by none other than Benjamin Franklin, and Mazzei was to form the Virginia Wine Company, whose stockholders included Jefferson and George Washington. Initially, Mazzei had some modest success, eventually producing two barrels of wine using multiple varieties of grapes, but this success was far from what he needed for financial success. Nonetheless, he was still confident about the conditions in Virginia and wrote to George Washington in 1779 that "experience has convinced me, that this Country is better calculated than any other I am acquainted with for the produce of wine."[51]

The American Revolution soon derailed Jefferson's and Mazzei's viticultural ambition, however. After the war, Jefferson attempted to grow grapes again, but still was not able to produce any significant amount of consumable wine. To complicate matters even further, many of his "European" vines were decimated by phylloxera (a louse or aphid that eats the roots of grape vines, often killing them). Native North American vines were largely immune to phylloxera, but Jefferson's European vines perished.

Jefferson never succeeded as a wine producer but remained a passionate connoisseur of wine throughout his life. Eight years prior to his death, in 1818, Jefferson stated that "in nothing have the habits of the palate more decisive influence than in our relish of wines."[52] Jefferson's passion for wine and his hopes to begin a wine industry in Virginia would eventually influence the rise of a major American (and Virginian) wine industry.

Benjamin Franklin was another signer of the Declaration of Independence who was quite fond of certain fermented beverages. He is often attributed with saying, "God made beer because he loves us and wants us to be happy," but in fact, this is a misquote. The actual quote does not praise beer at all. It praises wine. In a letter written to the Abbé Morellet in 1779, Franklin ponders Biblical references to wine when he wrote, "Behold the rain which descends from heaven upon our vineyards; there it enters the roots of the vines, to be changed into wine; a constant proof that God loves us, and loves to see us happy."[53] So, to correct the paraphrased quote, God made us "wine" because he loves us.

## *Voyage of Discovery*

Ardent sprits were quite prevalent in Lewis and Clark's westward travels, despite the fact that Benjamin Rush, who was advising the expedition, provided the explorers with eleven rules to maintain proper health on the mission including two pertaining to alcohol: "The less spirit you use the better"[54] and "After having your feet much chilled it will be useful to wash them with a little spirit."[55]

While it is unknown how much alcohol was purchased and consumed (or how may "demon whiskey" foot baths took place) on the westward voyage, it is clear that drinks were not scarce on the Voyage of Discovery. A surviving receipt signed by Merriweather Lewis in the National Archives shows that they purchased 30 gallons of "strong wine," and as the expedition proceeded, there were numerous recorded occurrences of drunkenness. In a journal entry on September 14, 1803, [Lewis] noted that they "set out this morning at 11 oClock was prevented setting out earlier in consequence of two of my men getting drunk and absenting themselves. I f[i]nally found them and had them brought on board, so drunk that they were unable to help themselves."[56]

A few days later at the junction of the Ohio and Mississippi rivers, he noted that they had "landed on the Spanish side ... found a number of our men who had left camp contrary to instructions and drunk."[57]

Despite these problems—and the warnings of the venerable Dr. Rush—an alcohol-free voyage would never have been considered. The health benefits and therapeutic social and stress-relieving aspects of alcohol were essential to hold the expedition together. Without alcohol, it is doubtful that morale would have been maintained in an expedition such as the Voyage of Discovery.

## Shackleton's Whiskey

The necessity of copious amounts of booze on exploration voyages had not changed very much by the time of the Shackleton expeditions in the early twentieth century. A team from New Zealand made an interesting discovery in 2007 when they were in Antarctica to restore a hut that had been built by Sir Ernest Shackleton, the noted British explorer. To their surprise, as they began their restoration work, they uncovered three cases of Mackinlay's Rare Old Highland Malt Whisky. This whiskey had been deposited there during Shackleton's 1907–1909 Antarctic expedition aboard a ship called the *Ninrod* (a few years later in 1914, he mounted his more famous expedition aboard the *Endurance*). This expedition appears to have been anything but temperate with regards to the drinking habits of the crew. Shackleton was a strong advocate of provisioning his exploration vessels with a healthy stock of alcohol. In a later mission he stated that it was for "medicinal and celebratory purposes."[58] In fact, Shackleton brought a total of twenty-five cases of whiskey, twelve cases of brandy, and six cases of port for his crew and they made use of it. In Roland Huntford's book *Shackleton*, he describes an early Christmas celebration that the crew held in June 1908: "The men wore paper hats and funny noses; Alistair Mackay, the second surgeon, passed out after drinking two-thirds of a bottle of whisky."[59] Presumably the first surgeon remained conscious.

Whether the spirits were medicinal or not, one of Shackleton's most well-known quotes was, "If I had not some strength of will I would make a first class drunkard."[60]

Whether the abundance of whiskey helped or hindered their expedition, the mission failed to reach the South Pole. Shackleton had to turn around, finally returning to England in 1909.

Of the three cases discovered in Antarctica, one of them was sent to

Scottish whiskey producer Wyte & Mackey. After the century-old bottles were analyzed, they decided to create a replicated version of the whiskey that would be marketed as *Shackleton's Blended Malt Whiskey*. According to Vijay Mallya, an Indian billionaire who purchased Wyte & Mackey in 2007, "We're looking to get some excitement going about Scotch in the younger generation, and the Shackleton whisky is a huge marketing peg. Look at fashion—everything is going retro these days."[61]

## Temperance and Prohibition

Benjamin Rush's warnings regarding the dangers of "ardent spirits" may not have significantly impacted drinking habits during his own lifetime, but his words echoed throughout the nineteenth century and into the twentieth century. In *Ardent Spirits*, Rush not only details the dangers of excessively consuming spirits, but also outlines a strategy for society to follow. He states, "Let good men of every class unite and besiege the general and state governments, with petitions to limit the number of taverns."[62] Rush also advised heavy duties on spirits and called on Christian churches to take a lead in this battle against drunkenness. In fact, it was Christian churches that led the temperance movement that eventually led to Prohibition in the first part of the twentieth century.

A turning point for the temperance movement was the Second Great Awakening, a Protestant religious revival that began in the late eighteenth century and lasted until the mid-nineteenth century. This religious movement emboldened anti-alcohol sentiments from churches and religious organizations around the country. Temperance movements grew and evolved throughout the nineteenth century so that there were eventually hundreds of temperance groups around the country—and they did have an impact. The annual average consumption dropped to 3.5 gallons between the years 1840 and 1860, dramatically down from the 7.1 gallons annual consumption in the 1830s. Support for complete Prohibition also grew toward the end of the nineteenth century, to the point where backing for the illegalization of all alcoholic beverages eclipsed that of mere temperance. Groups like the Women's Christian Temperance Union and the Anti-Saloon League had been extremely successful in advancing this cause and gained substantial political influence by the early twentieth century. As such, they were well positioned to act when national and international affairs provided them with an opportunity to achieve victory for their cause.

This opportunity came with the American entry into World War I in 1917. To increase grain production for the war effort, President Woodrow

Wilson ordered a temporary prohibition of alcohol in 1917. Even though this was not intended to be a permanent measure, Prohibition was now status quo and it did not take long for permanent Prohibition to take hold. During that same year, the 18th Amendment that called for the illegalization of alcoholic beverages passed both houses of Congress.

There were other factors that helped secure victory for the Prohibition movement. Alliances were made with the women's suffrage movement, providing additional momentum for this cause. In addition, passage was also influenced by anti–German propaganda by such groups as the Anti-Saloon League (ACL). According to the Prohibition website PBS, "ASL propaganda effectively connected beer and brewers with Germans and treason in the public mind. Most politicians dared not defy the ASL and in 1917 the 18th amendment sailed through both houses of Congress; it was ratified by the states in just 13 months."[63]

Christian Protestant groups may have led the temperance cause, but other religious communities were opposed to Prohibition. While the Jewish community in the United States had a reputation for being a "sober people," they tended to be opposed to Prohibition. One of the reasons for this described in Marni Davis's *Jews and Booze* was that they positioned themselves as "staunch defenders of the Constitution and champions of religious pluralism and political liberalism."[64] Another possible reason for their stand was that Jewish merchants were heavily involved in the alcohol importation and distribution industry and alcohol, specifically wine, was a key part of their religious observances.

Implementation of the 18th Amendment was famously complicated, triggering the raucous era known as the "Roaring Twenties." Speakeasies and bootlegging became synonymous in this era. However, bootlegging came from some unusual sources, including the Catholic Church. Sacramental wine is an essential part of the performance of the Eucharist. At first, even sacramental wine was banned during Prohibition, but in 1922, an exception was made for the use of sacramental wine used in religious ceremonies. According to an article that appeared on Winepair.com, "First, wineries had to obtain permits from the Prohibition director. Then a religious leader had to act as the proprietor of the winery when it came to production and distribution, and the same leader had to ensure that the wines were used for religious purposes, not general consumption. Under no circumstances could wine be consumed at the wineries."[65]

Despite these tight restrictions, grape production in California increased by 700 percent during Prohibition. The Department of Research and Education of the Federal Council of the Churches of Christ reported in 1925 that "the withdrawal of wine on permit from bonded warehouses for sacramental purposes amounted in round figures to 2,139,000 gallons in the fiscal year 1922; 2,503,500 gallons in 1923; and 2,944,700 gallons in 1924. There is no way of knowing what the legitimate consumption of fermented sacramental wine is but it is clear that the legitimate demand does not increase 800,000 gallons in two years."[66] This loophole also helped sustain the American wine industry.

There were even accounts that Catholic priests became illegal bootleggers. An example of this came from American author William Faulkner, who claimed that his bootlegger was a Catholic priest in New Orleans who took booze orders from the belfry of the St. Louis Cathedral.

Since wine was an essential part of Jewish religious rituals, Jewish merchants also entered into the sacramental wine business, with shops displaying wine in their storefronts with signs stating, "Kosher Wine: For Sacramental Purposes."

The 18th Amendment remained the law of the land throughout the 1920s until it was officially repealed on December 5, 1933, early in the first term of President Franklin Delano Roosevelt—and the timing for the repeal of Prohibition seemed appropriate. It was the height of the Great Depression, and when it was finally repealed, Franklin Delano Roosevelt reportedly toasted "What America needs now is a drink."

## Prohibition in France?

The United States was not the only country that faced a temperance movement in the early twentieth century. As mentioned earlier, Russia had entered into a period of Prohibition during the war years and there were temperance and Prohibition movements across the world—even France. In France, however, it was primarily focused on a specific drink—Absinthe. Absinthe, also known as "the green fairy," is an anise-flavored spirit that has a high alcoholic content that grew in popularity in the nineteenth century during a phylloxera outbreak in France that was devastating the wine industry. "The green fairy" became particularly "hip" in French society and was a favorite drink of such artists and writers as Edgar Degas, Edouard Manet, Paul Verlaine, and Vincent Van Gogh. According to Marie-Claude Delahaye of the Museum of Absinthe in Auvers-sur-Oise, "Absinthe was the queen of the Parisian boulevards."[67]

Absinthe had a reputation of having mind-altering effects and causing madness. Ironically, the French wine industry had a vested interest in this particular Prohibition movement. As they recovered from the phylloxera outbreak, wine producers wanted to eliminate the competition that had arisen during the years they had been sidelined. However, they needed to be very careful to guarantee that any law that was established was specifically focused at absinthe but did not hinder or prohibit wine sales. Whether or not the well-publicized deleterious effects of absinthe were true, the anti-absinthe forces were successful and "the green fairy" was banned in 1915. While there was some pressure to extend the Prohibition beyond absinthe, they failed.

## Bar/Pub Culture

> A rabbi, a priest, and a Lutheran minister walk into a bar. The bartender looks up and says, "Is this a joke?"

Bar room anecdotes and dramatic scenes have been a staple of popular culture for centuries. The cultural dynamics of bars and pubs are often not adequately factored into debates regarding the risks and merits of alcohol. Current debates often center around studies regarding the cardiovascular benefits of wine, or cancer risks from alcohol, or substance abuse. But these studies often ignore other major variables that make both the benefits and risks of alcohol far more complex. Bar/pub culture has played an important role in social interactions, business, romance, politics, and sometimes nefarious activities. According to Jason Harrell, author of *The Bartending Therapist*, a "bar is structured like a church. People congregate for many reasons—socializing being the most common…. Just like taking your seat in a pew, you feel at home…. Your attention is focused, and the bartender, much like a preacher, guides your practice of worship…."[68]

This is particularly true in the United Kingdom and Ireland where "the pub"—or "public house"—remains a key center of local public life. People from every walk of life come together to share a pint of a local brew, socialize, and relax after the work day. A recent Oxford study supports the importance of local pubs. According to Oxford Professor Robin Dunbar, "This study showed that frequenting a local pub can directly affect peoples' social network size and how engaged they are with their local community, which in turn can affect how satisfied they feel in life." He goes on to say that, "while pubs traditionally have a role as a place for community socializing, alcohol's role appears to be in triggering the endorphin system, which promotes social bonding."[69]

Unlike bars in the United States, children are allowed in pubs and it is not unusual for entire families to frequent these establishments, where they serve as hubs of community socializing and engagement from an early age.

Some American bars still serve similar roles, but they are not as family friendly. Perhaps the most famous fictitious bar in recent years appeared in the 1980s sitcom, *Cheers*. As proclaimed in the opening theme song, *Cheers* was a bar "where everybody knows your name." It represented an American version of the neighborhood bar similar to the British pub.

These types of bars still do exist in the United States, but they are often the exception rather than the rule. Many bars thrive on anonymity, which is the antithesis of English pubs. This solitude can certainly be a destructive force if the sole purpose of an individual is to anonymously get drunk, but it can also allow people to find solitude to ponder their problems or decompress without the stress of social interaction. Social interaction is often a very positive thing, but everyone occasionally has a need to escape questions and interaction of friends, family, and colleagues.

## The Therapist Bartender

> *"Oh, you hate your job? Why didn't you say so? There's a support group for that—it's called everybody, and they meet regularly at the bar."*
> —Drew Carey, *The Drew Carey Show*

At the center of the American bar mythology is the therapist bartender. The concept of a "therapist bartender" has almost become a cliché in popular culture, but they actually do exist. While bartenders should never be considered adequate replacements for psychoanalytical professionals, the bartender's role can have a positive impact on patrons—beyond doling out beer, spirits, and wine. A bartender does, however, have at least one advantage over a professional therapist. No matter how skilled a therapist may be, the vast majority of people—even ones who honestly seek counseling—build up emotional and psychological walls that remain in a therapist's office. However, after a couple of drinks, most individuals begin to shed some of these barriers and allow a more honest conversation. As psychologist David Skinner states, "When a person is drinking, the alcohol disinhibits them, so they start talking about their marriages or they start talking about their hopes or their dreams because they simply are not as concerned about other people's judgments."[70]

This is not always a positive development for the bar patron or the bartender, but it provides an accomplished "therapist bartender" an opportunity that a professional therapist does not have to penetrate some of these defenses.

Part of this rapport comes from the fact that a bartender will relate to his "client" as a friend whereas a therapist or psychologist relates from a position of authority—specifically not as a "friend." In addition to listening, a key attribute to be successful in this role is to be a generalist, well-read in many areas of current events and knowledge. As recalled by bartender Jason Harrell, "One day I am discussing Elon Musk and Steven Hawking to a SpaceX engineer, and the next day I will talk about the latest sporting acquisition of the LA Lakers with a journalist from the *LA Times*."[71]

## Diplomacy and Civilization

Elegant embassy parties featuring tuxedoed spies consuming a never-ending flow of martinis are a staple of Hollywood fiction and spy novels. Most embassy parties do not have as much intrigue and excitement as in a James Bond movie, but they are one of the central intersections between international diplomacy and libations. Alcoholic beverages are synonymous with many business and political social gatherings around the world. They can help both to improve social relations between diplomats and sabotage these interactions if any of the players overindulge. Regardless of the pros and cons, they have been a part of diplomacy and international relations for centuries.

One of the most famous centers of barroom diplomacy is the North Delegates' Lounge at the United Nations Headquarters in New York City. Until a 2013 remodeling, it was described as "dark and smoky and filled with Barcelona chairs and white leather. It was 'out of James Bond....'"[72]

While the new design was received with mixed reviews—one U.N. employee described the new look as "like a terrible airport lounge."[73] It nevertheless continues to play the same role it has for decades. The Lounge is an extremely popular haunt for delegates, staff, and others from around the world to unwind, socialize, and conduct a more relaxed form of diplomacy. As described by *The New Republic*, "The United Nations has six official standing committees; the in-joke is that the lounge is its seventh, because of all the bilateral negotiating that gets done there pre–6 p.m."[74]

However, after 6:00 pm, this international watering hole has a reputation for becoming rowdy, so much so that the Lounge staff needs to switch to using plastic cups rather than glass to avoid an inordinate level of breakage.

Excesses undoubtably occur, and serving alcohol in some cultures is highly inappropriate, but nonetheless, off-hour gatherings at "diplomatic watering-holes" can provide a more relaxed environment to build relationships and make deals while saving face.

## Conclusion

It is an indisputable fact that alcohol has played a major role—both positive and negative—in human history. Traditions and norms for consuming alcoholic drinks have changed over the centuries, but alcohol remains embedded in society, and shows no signs of fading. Once humanity begins to create permanent settlements and colonies in space (or even sustainable tourism), intoxicating drinks and "the bar" will follow, not merely out of the desire to feel "tipsy" but also because of the deep psychological and sociological tradition that alcohol represents. In fact, astrophysicist Ian O'Neill believes that when we travel to far-away worlds, these traditions may become more important. "We take alcohol for granted on Earth, but in space, I suspect social and ritualistic uses of alcohol will become amplified. Imagine being the first expedition to climb Olympus Mons on Mars, the highest volcano in the solar system, without a beer or a shot of whiskey to celebrate the feat after reaching the peak ... the accomplishment may well have not happened!"[75]

The social, psychological, and even physiological connection between alcohol and humanity is highly complex. Alcohol is both a product of good and bad behavior; morality and debauchery; intelligence and ignorance. Perhaps Judge Noah "Soggy" Sweat described this best in his famous "Whiskey Speech" in 1952 when he said, "If when you say whiskey you mean the devil's brew, the poison scourge, the bloody monster, that defiles innocence, dethrones reason, destroys the home, creates misery and poverty ... then certainly I am against it. But; If when you say whiskey you mean the oil of conversation, the philosophic wine, the ale that is consumed when good fellows get together, that puts a song in their hearts and laughter on their lips, and the warm glow of contentment in their eyes; if you mean Christmas cheer ... if you mean that drink, the sale of which pours into our treasuries untold millions of dollars, which are used to provide tender care for our little crippled children, our blind, our deaf, our dumb, our pitiful aged and infirm; to build highways and hospitals and schools, then certainly I am for it...."[76]

One should not diminish the dark and often tragic side of alcohol consumption, but the substantial positive elements are often ignored. Evolutionary psychologist Dr. Robin Dunbar maintains that a large part of the success of human society can be directly attributed to alcohol, particularly because of its role in human social interaction. "It isn't just because alcohol causes people to lose their social inhibitions and become over-friendly with their drinking chums. Rather, the alcohol itself triggers the brain mechanism that is intimately involved in building and maintaining friendships in monkeys, apes, and humans."[77]

# 2

# Booze in Science Fiction

*"I told you. I'm Guinan. I tend bar and I listen"*
—*Star Trek: The Next Generation*

It would be a vast overstatement to imply that alcohol has been the primary ingredient for successful space-based science fiction. Nevertheless, alcoholic beverages and interplanetary saloons have played an important role in this genre in innumerable stories in literature, film, and television. Romulan Ale and Klingon Blood Wine are well known outside science fiction circles, and the Cantina bar in *Star Wars* is so pervasive in popular culture that a significant percentage of the population can identify the theme music for that space saloon.

Such examples, and dozens more, do not by themselves provide empirical evidence that consumable alcohol will be commonplace in the space settlements of the future, but science fiction does have a remarkable track record for accurately predicting future technologies and trends. In many cases this genre's writers have actually been more accurate than NASA and professional scientists at prognosticating the future. As with the bottle of Dom Pérignon floating through space in the opening scene of the movie *Star Trek Generations*, science fiction writers tend not even to question whether "drinks" will be a part of future spacefaring society. Fermented beverages have been a part of human culture since the beginning of recorded history (and probably before) and are not likely to go out of fashion at space settlements and colonies in years to come.

It is not surprising that the visionaries of science fiction are actually in a better position than government agencies to make realistic depictions of future societal habits. This is not because NASA and other space agencies lack imagination. It is just not their role; their role is to set ambitious goals that are achievable within a limited time period and budget. As such, it is impractical for them to plan too far out into the future or to speculate about what society might look like in a hypothetical space colony many decades in the future. When they do speculate, these visions tend not to delve into the more controversial or gritty elements of society. "Official" depictions of future colonies often have a utopian—almost antiseptic—feel to them. What they often lack is real humanity, including the rough edges, the highs and the lows, the anxiety and fear, sex, and drinking. All these elements, however, have played—and will likely continue to play—significant roles in human culture and interactions. Great technological and scientific achievements are absolutely necessary, but humanity is still *humanity* and the prospect of a utopian future that is alcohol-free is highly doubtful.

## A Timeline of (Fictional) Drinking in Space

The concept of alcoholic beverages in space is anything but a recent development. Science fiction has incorporated interplanetary bars and saloons, as well as exotic spirits and cocktails, in storytelling for many decades.

While H.G. Wells did not provide his characters with alcohol in his 1901 story, *The First Men in the Moon*, he did write of their craving for beer during their voyage. The lunar explorers were experiencing hunger and thirst, and the narrator, Mr. Bedford, described his growing appetite for food and drink, "My mind ran entirely on edible things, on the hissing profundity of summer drinks, more particularly I craved for beer. I was haunted by the memory of an eighteen-gallon cask that had swaggered in my Lympne cellar."[1]

No casks of lunar brew were to be found on the Moon, but they did find what Wells called "monstrous coralline growths"[2]—a fungus growing from the floor. Bedford's hunger gets the better of him and he eats some of the fungus. He then explains, "The stuff was not unlike a terrestrial mushroom, only it was much laxer in texture, and, as one swallowed it, it warmed the throat. At first we experienced a mere mechanical satisfaction in eating. Then our blood began to run warmer, and we tingled at the lips and fingers, and then new and slightly irrelevant ideas came bubbling up in our minds.... The depression of my hunger gave way to an irrational exhilaration."[3] Bedford and his companion Mr. Cavor became intoxicated by these lunar mushrooms. Their intoxication was unintentional, however. They were driven by hunger rather than a need for a lunar "bender."

Several decades earlier, Jules Verne, in his 1865 novel, *From Earth to the Moon* (translated to English in 1867), kept his fictional spacecraft well stocked with fermented beverages. As a Frenchman, it should not come as a surprise that Verne provided wine for his lunar explorers. As Verne's crew travels through space, getting closer and closer to the Moon, they dine and also speculate about winemaking on the lunar surface: "Nothing was so excellent as the soup liquefied by the heat of the gas; nothing better than the preserved meat. Some glasses of good French wine crowned the repast, causing Michel Ardan to remark that the lunar vines, warmed by that ardent sun, ought to distill even more generous wines; that is, if they existed. In any case, the far-seeing Frenchman had taken care not to forget in his collection some precious cuttings of the Médoc and Côte d'Or, upon which he founded his hopes."[4] This reference to winemaking on the Moon may well be the first time that the concept of winemaking on another planetary body was proposed. As will be outlined in a later chapter, Jules Verne was 150 years ahead of his time. There are now numerous projects around the world investigating whether wine and beer can be manufactured on the Moon and Mars.

More recently, Ray Bradbury did not imagine a "dry" planet in his novel, *The Martian Chronicles*. On the contrary, wine was quite abundant on the Mars of Bradbury's imagination. Not only did the Martians consume their own variety of wine, but wine appears to be a standard issue supply aboard the various spacecraft sent from Earth. Perhaps the most poignant example of this appears toward the end of the book, years after a devastating war on Earth. Mr. Hathaway and his family are living alone on Mars, and doubt whether they will ever see any other humans again, but one evening Hathaway sees the glow in the sky of a rocket beginning its descent to Mars. He rushes home and pulls out an old dusty bottle of wine he had held on to for decades. He declares to his family, "Wine I saved, just for tonight. I knew that someday someone would find us! We'll have a drink to celebrate!"[5]

The next day, the rocket arrives and Captain Wilder, an old colleague of Hathaway's, steps from the spacecraft and reveals that he and his crew had been exploring the outer planets for the last several decades, and had only recently become aware of the tragedy that had befallen both the Earth and Mars. Despite being away from Earth for so long, they were still able to retrieve a bottle of wine from their spaceship to celebrate their reunion. Hathaway raised his glass to his guests, stating, "A toast to all of you; it's good to be with friends again. And to my wife and children, without whom I couldn't have survived alone. It is only through their kindness in caring for me that I've lived on, waiting for your arrival."[6]

Bradbury clearly believed that wine and other beverages would be a standard supply for future colonists on Mars. While this will not likely be a priority (or even permitted) if NASA sends humans to Mars in the upcoming decades, it will almost certainly be a reality if private one-way missions ever become a reality.

## DIY Space Drinks

Not all stories of interplanetary imbibing have required the crews to stow booze from Earth aboard their spaceships. Indeed, many science fiction authors have predicted that alcohol will be manufactured locally even in the very early days of deep space exploration. For example, noted science fiction author and astrophysicist Gregory Benford wrote of Martian "home brews" in his 1988 short story, "All the Beer on Mars." In the not-so-distant future, an international crew is searching for life on Mars. During this search, a crew member named Lev successfully brews beer utilizing their food stores. "He had smuggled the yeast on the expedition and experimented with it during the eight months voyage. They recycled their water and the brewing concealed the processing tastes. It was the best possible morale booster in a world of stinging aridity."[7]

Consuming this beer becomes a core part of their social interactions, and when they confirm the existence of microbial life on Mars they raise their beakers of beer in celebration, toasting "Here's to life on Mars."[8] As their celebration continues, one of them proclaims, "Come on. We'll drink up all the beer on Mars."[9] In this way Benford creates a highly realistic scenario. It does not take a tremendous stretch of the imagination to conclude that the first explorers to Mars will find a way to celebrate with a drink. According to Benford, "I was basing it on the experience of many expeditions to hard parts of the world in the nineteenth century…. It's a human need. You're living in a completely hostile environment that tries to kill you every day and the human compensation for that is to distract yourself and elevate your mood—so it's medically needed."[10]

Benford is not alone in believing that beer will one day be produced in space. David Brin also anticipates brewing beer in space in his 2011 short story, "Tank Farm Dynamo." The story takes place on a space station constructed from expended fuel tanks from rockets launched from Earth. A private company utilizes these fuel tanks, as well as the unexpended fuel (liquid hydrogen and oxygen) that remains in them, to construct and operate a facility to "slingshot" spacecraft that have been launched from Earth into higher orbit.

However, assisting spacecraft to achieve higher orbits is not the only product of this facility. One of their crewmembers doubles as a brewmaster and creates Slingshot Beer using yeast, barley, and hops, all grown in sludge in an old tank that is used as a garden, as well as water that is manufactured using the unexpended liquid hydrogen and oxygen from the fuel tanks. Brin is unsure how viable actual fermentation would progress in microgravity, however, noting that "we don't know yet if fermentation is any different in space. Certainly, distillation will be more of a problem, without a sense of up and down."[11] Since the Tank Farm generates a small amount of gravity, Brin sidesteps the question of fermentation in weightlessness.

As Slingshot is the only beer produced in space, it becomes extremely popular among station visitors and becomes an unlikely yet compelling incentive for visiting the station. Regardless of the quality of this brew, the novelty of consuming such a unique product is extremely alluring to many individuals. According to one of the station visitors, "It costs a hundred bucks a pint on Earth. It's a damn fine beer."[12] Visitors to the Tank Farm do not have to pay for their beer, but $100 for such a pint on Earth would be quite reasonable in a real-world market. Collectors of rare wines, scotch, and other beverages will often pay tens of thousands of dollars—and even a lot more—for a rare or historic bottle of booze. For example, in 2010 eleven bottles of 170-year-old champagne that had been in a shipwreck at the bottom of the Baltic Sea were auctioned off. Auction sales totaled $156,000, with one bottle of Veuve Clicquot selling for 15,000 euros.

It should also be noted that Japanese beermaker Sapporo has sold a limited number of six-packs of their Space Barley beer (made from barley descended from plants that have flown in space) for $110. When the first beer is fully manufactured in space, that brew will unquestionably garner a much higher—perhaps astronautical—price.

## To Drink Where No One Has Drunk Before

> Kirk: "Romulan ale? Why Bones, you know this is illegal."
> McCoy: "I only use it for medicinal purposes."
> —Star Trek II: The Wrath of Khan

Some of the most famous fictional space-born drinking scenes appear not in the pages of science fiction novels, but on television and in motion pictures. Specifically, the *Star Trek* franchise has not shied away from alcohol consumption in their storylines. While this iconic franchise did not provide the first portrayals of drinking in space and did not depict the most constant scenes of imbibing in the annals of science fiction, *Star Trek* has arguably had more impact on the concept of "space drinks" than any other science fiction franchise or story.

There were numerous drinking scenes in the original series. One of the most memorable scenes occurred in the 1967 episode, "The Trouble with

Tribbles," in which a rapidly reproducing species known as Tribbles were introduced into the *Enterprise*—where they became an invasive species to the extreme. During the episode, Chief Engineer Montgomery "Scotty" Scott, Ensign Pavel Chekov, and another crew member enter into a drinking contest. Predictably, Scotty drinks scotch and Chekov imbibes vodka. However, this game is interrupted by a belligerent Klingon and a barroom brawl ensues. Meanwhile, Cyrano Jones, a guest character who was responsible for the Tribble infestation, uses the commotion to consume as many free drinks at the bar as he can before the ruckus dies down.

This was far from the only scene featuring the crew of the USS *Enterprise* becoming slightly or very inebriated. In a 1968 episode called "By Any Other Name," the ship encounters an intellectually superior species called the Kelvans who take over the ship using a device that can paralyze the crew. The alien intruders then begin to experience human senses for the first time. Kirk, McCoy, and Scotty brainstorm how to gain the upper hand on their unwanted guests and determine that they must steal one of these paralyzing devices. Since the Kelvans have no experience with any form of sensory pleasure such as taste and smell, Kirk proposes to take advantage of this inexperience and asks his officers to "look for any way to stimulate the senses." Scotty replies confidently, "I can think of one way right off."

Scotty approaches an alien captor who is sitting at a table, voraciously eating his first meal. Scotty asks, "Lad, you're going to need something to wash that down with. Have you ever tried any Saurian brandy?" The two of them then partake of the contents of three bottles, but Scotty's drinking foe still appears to be unaffected. Scotty then says sadly, "All I have left is a bottle of very, very, very old Scotch whisky!" Nonetheless, to save the ship, Scotty opens his prized, aged bottle. Finally, as the last precious drops of whiskey are about to be imbibed, the Kelvan captor falls to the ground, unconscious. Scotty then kisses the scotch bottle and says, "We did it, you and me. Put him right under the table."

To avoid the perception that members of Star Fleet were perpetually exploring the final frontier while under the influence, the producers of *Star Trek: The Next Generation* introduced the concept of "synthehol"—a beverage that is supposed to simulate the taste and effects of alcohol. Described by *Star Trek* creator Gene Roddenberry, synthehol "acts just the same as alcohol. It makes you feel that you can be a lover or wise or all the things that alcohol does, but it's only temporary. With force of will you can shove it aside and be as sober as you ever were...."[13] According to Jonathan Frakes, who played Commander William Riker, "I think Gene Roddenberry enjoyed the occasional cocktail and was unabashed about its value in his story-telling. But given that we were a group of Star Fleet officers, he essentially created a drink that we could enjoy while we drank it but was magically shaken off."[14]

Despite the existence of synthehol in the *Star Trek* universe, Star Fleet officers would still occasionally indulge in real drinks, even in later series like *The Next Generation, Deep Space Nine,* and *Voyager.* In *Star Trek, The Next Generation*, Captain Jean Luc Picard, portrayed by Patrick Stewart, comes from a family of winemakers in France. In a 1990 episode called "Family," Picard becomes inebriated along with his older brother as he (Captain Picard) tries to cope with the fact that he had been abducted and controlled by the Borg—the cybernetic bad guys in this franchise. While under their control, he helped to kill thousands of his Star Fleet colleagues. Only a drunken brawl with his brother allows him to release his anguish and horror at his previous actions and to learn to cope with that knowledge.

The ship's counselor, Deanna Troi, also succumbs to the lure of inebriating beverages. In the motion picture *Star Trek: First Contact*, Troi overconsumes at a grungy bar while searching for Zefram Cochrane, the inventor of warp drive. After Commander Riker inquires whether she is drunk, she replies, "I am not. Look, he wouldn't even talk to me unless I had a drink with him. And then it took three shots of something called Tequila just to find out he was the one we're looking for."

In the *Star Trek* universe, even androids are not immune from the pleasures of a libation. After receiving an emotion chip, Lieutenant Commander Data is anxious to fully experience the new world he is being exposed to—one where he not only feels emotions, but also can appreciate all of his new senses. This includes having a drink. At the *Enterprise* bar Ten Forward, Data gulps down a beverage that ship's bartender Guinan calls "something new from Frokas III." Data gags and announces, "Ah yes, I hate this! It is revolting." Guinan asks, "More?," to which Data replies, "Please!"

It should not come as a surprise that there is a market for *Star Trek* themed alcoholic beverages and one innovative company is producing drinks to suit *Star Trek* fans who also appreciate fine spirits. Silver Screen Bottling Company, a business that works with film studios, actors, and others in the film industry to produce film-based/themed alcoholic beverages, offers several *Star Trek*-inspired spirits. These include James T. Kirk Bourbon (William Shatner is a limited partner in the company), Montgomery Scott Scotch, and most recently, Ten Forward Vodka. However, with Ten Forward they took the additional step of sending a bottle to the edge of space on a high-altitude balloon in 2018. According to Silver Screen Bottling sommelier Ryan McElveen, "We got some great images and then when we brought the bottle back down, we took it and blended it into the stocks of the rest of our bottles of vodka. Theoretically, every bottle in perpetuity will have a small portion of that bottle that has been to space."[15]

McElveen concedes that this was primarily a marketing gimmick, but it has been extremely successful and Star Trek fans have "totally gravitated toward it."[16]

Silver Screen Bottling has bigger ambitions for the future. McElveen explained, "We were originally going to put it on one of Elon Musk's projects. We were in talks about doing it. The reason we were going to do it with SpaceX is that Elon Musk happens to be a big *Star Trek* fan. We think that eventually, once he has a little less on his plate, we can start having a conversation in earnest and we may be able to get it up on one of his projects...."[17]

## *Interstellar Drinking Holes*

As shown in many of these scenes (as well as in the earlier historical chapter), bars, saloons, and pubs are the centerpiece of alcohol culture and tradition. Bars have served myriad roles in human society, well beyond merely being an outlet to "have a drink." There is a universality and an inevitability to the prospect of these gathering places far away from Earth. Indeed, even those who are not fans of science fiction can often identify fictional saloons in space. Perhaps the most famous intergalactic bar scene in the annals of science fiction is the Mos Eisley Cantina in *Star Wars: A New Hope*. Residing at the Mos Eisley spaceport, a location that Obi-Wan Kenobi describes as a "wrenched hive of scum and villainy," the Cantina has a thriving business catering to the villainous scum referred to by Kenobi. It is filled with galactic outlaws, smugglers, and other undesirable beings. However, we quickly learn that they do not serve alcohol to androids. As Luke is about to enter with C3PO and R2D2, the gruff barman yells, "Hey, we don't serve their kind here!"

The cantina bar scene only lasts for a few minutes, but plays an essential role in the plot development, providing a vehicle to viably bring the key protagonists together. Han Solo and Chewbacca specialize in smuggling, so the Cantina serves as a perfect location for them to acquire new semi-legal or illegal work.

The space bar has appeared in many science fiction tales over the years. According to John Spencer, the co-founder of Mars World Enterprises, "In TV shows and movies, where a lot of the intimate character dialog and interactions, and the revealing of things, happen in space bars. That is what happens in the real world."[18]

The Mos Eisley Cantina is so popular that a "pop-up" bar called Scum & Villainy Cantina appeared in 2016 in Los Angeles and became a permanent bar later that year. Scum & Villainy co-founder J.C. Reifenberg was inspired by the fact that there is a sports bar on the corner of practically every town in America, where sports fans dress up in sports apparel and discuss player and game statistics. "It started to sound really familiar—of people discussing movies and pop culture and nerd things. The fact that I know that the odds of successfully navigating an asteroid field are 3720 to 1 isn't all that different—in theory—than someone being able to pull up the fact that Joe Montana has × number of passing yards...."[19] Reifenberg thought the timing was perfect for such an establishment, since the more traditional gathering places of "nerds"—comic book stores—have been struggling in recent years. "Scum and Villainy is like a comic book shop that sells alcohol and food instead of comic books.... A place where you can go to put on your cosplay mask so that you can take off your [workplace] mask and be who you really are."[20] Essentially, it is a safe space for this community to gather to discuss their passion.

However, it is not just "nerds" who come to the Scum & Villainy Cantina. Reifenberg stated that people from all walks of life come in for the experience as well as a steadily growing list of actors, filmmakers, writers, and many others who have worked on some of the most important science fiction television and film productions.

Reifenberg recalled that his friend, filmmaker Kevin Smith, might have

Based on a drink that was featured on the television show *Battlestar Galactica*, ambrosia is just one of many science fiction-inspired drinks at the Scum & Villainy Cantina in Hollywood, California (Scum & Villainy Cantina).

described the role of Scum & Villainy best when he said, "Scum & Villainy and you are building a spot that brings people together. That's why it's important. It's not the product that you're putting out in the bar. It's the fact that you are giving people who don't have a place to go—a place to go."[21]

*   *   *

Probably second only to Mos Eisley Cantina in recognizability within the popular consciousness is Ten Forward of *Star Trek: The Next Generation*. Ten Forward became a primary set on the series, subsidiary only to the bridge and Engineering, but only fully developed as a critical set when show producers introduced a new bartender named Guinan, played by Whoopi Goldberg. According to Jonathan Frakes, Roddenberry loosely based Guinan on a real-life woman named Texas Guinan who was famous for running speakeasies in the 1920s. Frakes added that with the "social elements of a bar on a space ship, it blossomed for all of us. It was a place where people could go to unwind and mix—we had music in there. Everything about Whoopi and the bar enhanced *Next Generation*."[22] (Note: According to Mars World creator, John Spencer, Ten Forward was not just a fictional bar. *Star Trek* creator Gene Rodenberry actually had the set of Ten Forward fitted out to be a real-life bar. Studio executives would frequently run events there and bring in VIPs for drinks. Outside groups would even occasionally utilize it for fundraisers and other events.)

Guinan was formed very much in the "therapist bartender" model. She would listen to the problems of the crew and would dole out words of comfort, wisdom, and sometimes blunt chastisement to crewmembers who were struggling emotionally or losing confidence. Her blunt authority is clear in an episode called, "The Best of Both Worlds (Part 2)," after Captain Picard is kidnapped by the Borg. Commander Riker is now in command of the *Enterprise* and has lost confidence that he and his crew can prevail against the seemingly invincible Borg forces. Guinan confidently walks into his office and says, "I've heard a lot of people talking down in Ten Forward. They expect to be dead in the next day or so. They trust you. They like you. But they don't believe anyone can save them." Riker replies, "I'm not sure anyone can." Guinan responds by saying, "When a man is convinced he's going to die tomorrow, he'll probably find a way to make it happen." She goes on to give Riker the emotional boost that he needs to move forward.

Unlike Counselor Deanna Troi, who is the full-time therapist on the ship, Guinan has the advantage of not being part of the Star Fleet chain of command. Guinan is able to talk to crew members when they are more relaxed, when they let down their guard.

*"Don't call me barkeep! I'm not a barkeep! I'm your host, the proprietor, a sympathetic ear to the wretched souls who pass through these portals."*
—Quark, *Star Trek: Deep Space Nine*

The next series in the *Star Trek* franchise, *Deep Space Nine*, also featured a prominent bar. However, this bar represented the seedier elements of humanity as well as aliens who passed through that space station. It was more reminiscent of Mos Eisley Cantina than it was of Ten Forward. Quark's bar was owned by (and named after) Quark, a Ferengi businessman/operator, who not only ran a drinking establishment that featured gambling, female dancers of many species, and pornographic "holosuites," but also conducted numerous under-the-table business dealings. Like Ten Forward, Quark's bar was also one of the central locations for plot and character development, and provided an ample share of tension between Quark and Deep Space Nine Constable Odo.

Note: Quark's was even recreated in a theme hotel in Las Vegas. *Star Trek: Experience* was built at the Las Vegas Hilton in 1998 and featured a Promenade like *Deep Space Nine* as well as a bar modeled on Quark's bar. Unfortunately, after ten years of operation, it closed its doors in 2008.

Alcohol and synthehol consumption were less pervasive in *Star Trek Voyager*, but nonetheless alcohol still was used as a plot device—particularly to explore different characters. In an episode called "Body and Soul" during the seventh season of *Voyager*, the Holographic Doctor, played by Robert Picardo, downloads his consciousness to the cybernetically altered brain of crewmember (and former Borg) Seven of Nine, a role that is played by Jeri Ryan. In doing so, the Doctor's personality takes over Seven's body, enabling him to understand what it is to be human for the first time (similar to when Lt. Commander Data receives his emotion chip). Via Seven of Nine's body, the Doctor overindulges in both food and alcohol. According to Robert Picardo, "Jeri Ryan had to imitate my performance, which she did brilliantly … and she has a love affair with a piece of cheese cake … and after having a drink, the Doctor marvels and muses at the sensations of being human."[23]

In contrast to *Star Trek*, the 1990s science fiction soap opera *Babylon 5* depicted a grittier perspective of a future spacefaring humanity. Show creator J. Michael Straczynski portrayed a human race that retains all of the "warts" and vices that we have today. A five-mile-long space station in neutral space, *Babylon 5* contains numerous saloons that are an essential part of socialization, diplomacy, commerce, and the darker elements of that society. It is clear that this universe is a drinking culture, but it is not always portrayed in a positive light and the show did not shy away from depicting the destructive aspects of excessive alcohol consumption. In fact, the Security Chief, Michael Garibaldi (played by Jerry Doyle), is a recovering alcoholic. This theme remains a continuous thread throughout the series, as Mr. Garibaldi struggles (and sometime fails) to remain sober. This role was quite personal for Jerry Doyle. In his real life, Doyle struggled with an alcohol problem until his death in 2016.

Addiction was a consistent theme in *Babylon 5*. Even the head of medical operations, Dr. Stephen Franklin (played by Richard Biggs) becomes addicted to stimulants. Show creator J. Michael Straczynski's own father was a violent alcoholic who he described as "the most evil man I've ever known."[24] It is clear that these personal stories were drawn upon to add depth and develop plots that were genuine. Years later, in a 2018 tweet, however, Straczynski made it clear that while he acknowledged the reality of drinking, he did not want to be perceived as advocating for it. "Many people drink, it's what they do, so I include that as a natural part of the writing but coming from three generations of drunks and alcoholics I don't endorse it or have any interest in 'drinking culture.'"[25]

Despite this real-life backstory, the positive interactions at the bars of *Babylon 5* served as a core driver of the storylines throughout the series. Some of the more memorable drinking scenes occur between Ambassador Londo Mollari and Ambassador G'Kar and symbolize the changes in their relationship. They share a drink just before their worlds entered a devastating war and they later share a drink to sanctify a proposed interstellar alliance between their two peoples.

In fact, there were very few episodes that did *not* feature some form of drinking. A large number of these scenes involved Ambassador Mollari, who was portrayed as a prodigious drinker whose imbibing abilities would have put Winston Churchill to shame. Mollari explains his philosophy of drinking when he asks a visiting representative from Earth's government if he would like a drink, which his guest declines because he's on duty. Londo responds by explaining, "Well, you see, we Centauri are always on duty. Duty to the Republic, to our houses, to one another, and so we have made the practice of joy another duty. One that must be pursued as vigorously as the others. You should try it some time." Mollari then drinks from his glass and continues with business.

As shown in the 2016 motion picture *Passengers*, the bar dynamic does not require a large number of people to be effective. During most of the film, there were only two "human" characters who spend much of their time at the Concourse Bar interacting with an android bartender named Arthur. The bar served as a central location for character and plot development and was a crucible for emotional progression and conflict. According to Mars World co-founder John Spencer, who consulted on this film, "The bar is a great place for these characters to have intimate scenes and the audience feels comfortable because they have done it themselves."[26]

In the film, a mechanic named Jim Preston, played by Chris Pratt, is on a 120-year trip to a new world. Jim is revived from his hibernation pod prematurely as a result of a ship malfunction. He realizes that he is the only person awake with 90 years remaining. Fortunately, he bonds with the android bartender, Arthur, who is played by actor Michael Sheen. In these conversations, he struggles emotionally regarding whether he should wake up Aurora Lane, played by Jennifer Lawrence, who he has become infatuated with. To wake her up would assure that she did not live long enough to reach their final destination—and she would certainly not be pleased to learn that he had intentionally roused her from hibernation. Extreme loneliness overtakes rational thought, however, and he decides to wake her up, but claims it was caused by another mechanical failure.

Later in the film the role of the Concourse Bar comes full circle—in the most emotionally charged scene of the film—when Arthur informs Aurora that she had not been awakened by a malfunction of her hibernation pod, but that she had been deliberately revived by Jim Preston.

After this revelation, the two characters alternate "bar nights" and are careful to avoid each other until other plot developments force them together. *Passengers* effectively shows that the barroom device is effective in science fiction (and general fiction) even if there is only a bartender and one or two other people.

Science fiction novelist and futurist David Brin explains this well, stating, "Stories (and human life) turn much less around action or technology than conversation, which lets your characters express feelings, needs, anguish, joy and conflict. Well, nothing loosens the tongue [better] than several libations in a good bar. Which is why some of the best tall-tales in science fiction are set at sites like [Arthur C.] Clarke's The White Hart or Spider Robinson's Callahan's."[27] *Callahan's Crosstime Saloon*, published in 1977, is not technically a space bar, but it is a setting that Robinson uses to bring together multiple science fiction genres into one location—which happens to be a saloon. The saloon is owned and operated by Mike Callahan and provides a place where an unusual mix of clientele can drink, express their feelings or concerns, tell their stories, or remain silent. These patrons include extraterrestrials, time travelers, a talking dog, a vampire, and many other unusual guests sipping Irish whiskey. A guiding philosophy for this (quantum) drinking hole is "Callahan's Law" or the "Law of Conservation of Pain and Joy," which states that "shared pain is lessened; shared joy, increased—thus do we refute entropy."[28]

Gregory Benford wrote a story reminiscent of *Callahan's Crosstime Saloon* in his short story, "The Fairhope Alien." The story's premise is that a race of aliens called The Alphas come to Fairhope, Alabama, but rather than conveying some profound wisdom to the human race or asking the residents to "take us to your leader," one of the first things they do is stop at the local pub. As the narrator explains, they sit down at the bar and "they are ordering up and putting them down pretty quick. Nobody knows their chemistry but they must like something in gimlets and fireballs and twofers, cause they sure squirt them down quick."[29]

According to Benford, "the pub" may be a universally familiar location, stating, "Sentient creatures know what alcohol provides—a complex reaction that soothes your fears—and takes away the skittery anxiety that being intelligent means."[30] This was certainly the case in Douglas Adams' novel, *The Hitchhiker's Guide to the Galaxy*. In the first few pages of this volume, Adams describes an alien named Ford Prefect who had been living on Earth for fifteen years as "an eccentric, but a harmless one—an unruly boozer with some oddish habits. For instance, he would often gate-crash university parties, get badly drunk and start making fun of any astrophysicist he could find till he got thrown out."[31] As such, it only seems natural that when Prefect confronts Arthur Dent (who is from Earth) to warn him that the world will be ending in twelve minutes, he urges Dent to join him for a stiff drink at the local pub.

However, after a copious number of drinks in a short span of time and just as the Earth is about to be destroyed, Prefect and Dent hitchhike their way off the planet. Dent will soon learn that his final drinks on Earth need not be his last. He enters a bizarre galaxy where the various aliens really like alcohol. In fact, *The Hitchhiker's Guide*, "which has supplanted the great Encyclopedia Galactica as the standard of all knowledge and wisdom,"[32] reveals that the Pan Galactic Gargle Blaster is the best drink in the galaxy. Drinking it "is like having your brains smashed out by a slice of lemon wrapped around a large gold brick."[33] The guide gives mixology instructions as well as "what voluntary organizations exist to help you rehabilitate afterward."[34]

A more recent entry in the genre of interplanetary saloons comes in Andy Weir's novel *Artemis,* which tells the story of the first colony on the Moon. The protagonist of this lunar drama is a young woman named Jazz, who is a "sneaky criminal trying to get ahead. She was sort of a delinquent growing up, and now she realizes that she made a lot of mistakes in her youth and she's trying to make up for some of them. She is very flawed in a lot of ways, but hopefully also very likable."[35] Jazz frequents Hartnell's Pub, which Weir describes as "a hole in the wall. No music. No dance floor. Just a bar and a few uneven tables.... Hartnell's was for drinking. And you could get any drink you wanted as long as it was beer."[36] This is not the case in all areas of the Artemis colony. According to Weir, "Artemis is a vacation destination. So, they import alcohol from Earth and sell it at a huge markup. But they're still in an era where sending mass from Earth isn't cheap. So, they tend to have reconstituted beer and spirits..."[37] This is precisely what Hartnell's bartender, Billy, has in mind and he hopes to recreate more fancy drinks for his bar. In one case, he offers Jazz what he indicates is "Bowmore single-malt scotch,"[38] causing Jazz to spit it out. Seeing Jazz's reaction, Billy says, "I had a bloke on Earth boil the liquids off then send me the extract. I reconstituted it with water and effanol. Should be exactly the same."[39] While real scotch is rare on Artemis, Jazz is familiar with the taste of the genuine product through one of her wealthy customers who would spend large sums of money to have it shipped from Earth. As such, she knows instantly that Billy's scotch experiment has gone badly wrong.

The fact that there is alcohol at all at the Artemis station is a compromise to social reality. Since ethanol is a flammable substance, pure ethanol is restricted, but as Weir explains, "In a concession to human nature, Artemis allows liquor even though it's flammable."[40] As will be described later in this book, this concession is highly realistic. In present-day space exploration, consumption of alcohol is "officially" banned in space by space agencies around the world. The flammable nature of ethanol is often cited as well as the fact that the fumes can have an adverse effect on systems. That said, unofficially, it is well known that certain concessions are allowed regarding alcohol in space.

Whether they are frequented by aliens or time travelers is impossible to predict, but space-based bars will almost certainly become reality if humans begin to settle other parts of the solar system or galaxy. Astrophysicist Ian O'Neill believes space bars are inevitable. "It'll most likely start small, perhaps a few Mars inhabitants setting up a micro-distillery, creating a catalyst for the first 'Mars bar' ... as we're only talking about a few dozen/hundred/thousand explorers, it will probably be in the style of a prohibition-era 'speakeasy' ... I always imagine *Battlestar Galactica*'s 'Joe's Bar' as being the quintessential first-colony Mars Bar ... it has the unofficial blessing of the commanding officers, all while becoming an invaluable component of the social structure...."[41]

## Famous Space Bars

Interplanetary watering holes represent some of the most memorable scenes in the annals of science fiction. While it would not be practical to attempt a comprehensive list of these fictional off-world drinking establishments within the pages of this book, just a few of the more notable ones not yet mentioned include (in no particular order):

• *Joe's Bar*: Joe's Bar appears on the reboot of *Battlestar Galactica* (2004–2009). Hidden on the flight decks of *Galactica*, Joe's boasts a pool table, piano, and a Viper (a fighter spacecraft) hanging from the ceiling similar to what one might find in an aviation bar in the United States. The bar was introduced to the series in the third season in an episode called "Taking a Break from All Your Worries," a line borrowed from the theme song of *Cheers.*

• *The Royale*: The SyFy series, *Killjoys*, featured The Royale, a saloon located in Old Town on Westerley. It's owned by a former warlord named Pree, who now focuses his attention on the pleasure and satisfaction of his clientele in the run-down Old Town.

• *Taffey's Snake Pit Bar*: Taffey's, featured in the 1982 film *Blade Runner*, has a similar ambiance to that of the Mos Eisley Cantina. Owned by Taffey Lewis, Taffey's is a smoke-filled bar with exotic music, drinks, and dancers on display. *The Blade Runner* film notes describe it as "a decadent bar in a rough part of town catering to the wealthy."

• *Cowboy Feng's Space Bar and Grill*: Like the Tardis in *Doctor Who*, the bar and grill in Steven Brust's 1990 novel is frequently transported through space and time. The occupants of the bar have no control over these shifts in time and space. The catalyst for these moves is when the city where the bar currently resides gets obliterated by nuclear war. Nuclear war seems to be rampant in this universe since Cowboy Feng's had traveled from Earth, to the Moon, to Mars, and many other destinations just as nuclear explosions decimate their previous homes.

• *Milliways*: Milliways is a restaurant that appears in *The Restaurant At the End of the Universe* that literally exists at the end of time and space. As explained by Milliways' Max Quordlepleen, "Ladies and gentlemen, the Universe as we know it has now been in existence for over one hundred and seventy thousand million billion years and will be ending in a little over half an hour. So, welcome one and all to Milliways, the Restaurant at the End of the Universe!"[42]

# Alien Drinks

Like space bars, the space and alien drinks of science fiction are far too numerous to list in this book, but a few of the better-known interplanetary libations include:

- Romulan Ale: In *Star Trek*, this extremely potent Romulan alcoholic beverage was illegal within the United Federation of Planets. Nonetheless, it did not stop rambunctious and somewhat rebellious Star Fleet officers from occasionally imbibing. In one memorable scene in *Star Trek: The Undiscovered Country*, Captain Kirk and other crew members have a tense dinner party with a delegation of Klingons and sip Romulan ale while spouting Shakespeare in both English and in "the original Klingon."
- Klingon Blood Wine: Another *Star Trek* alien drink is blood wine. This highly intoxicating Klingon beverage is used in rites of passage and other ceremonies and is an acquired taste even for Klingons. Incidentally, Votto Wines released a Malbec, Syrah, Petit Verdot blend called "Klingon Bloodwine" in 2014. No actual blood was added to this wine.
- Pan Galactic Gargle Blaster: This drink is from *The Hitchhiker's Guide to the Galaxy*, which is described as "like having your brains smashed out by

Based on a drink in *The Hitchhiker's Guide to the Galaxy*, the Pan Galactic Gargle Blaster is just one of many science fiction drinks inspired at the Scum & Villainy Cantina in Hollywood, California (Scum & Villainy Cantina).

a slice of lemon, wrapped 'round a large brick.'"[43] This alien concoction is also described as the best drink in existence.

• Janx Spirit: Another drink described in *The Hitchhiker's Guide to the Galaxy*, this drink was often used in a drinking game played in the Orion Beta star system. Two players would sit opposite each other and "concentrate their will on the bottle and attempt to tip it and pour spirit into the glass of his opponent, who would have to drink it,"[44] hindering the opponent's "telepsychic" power.

• Space Melange, otherwise known as "the Spice," in Frank Herbert's novel *Dune*. The Spice has mind-altering effects, is extremely addictive, can change a drinker's physical appearance, and can increase life expectancy.

• Jovian Sunspot: Based in the *Babylon 5* universe, this is a cocktail created by humans that is popular with Babylon 5 officers, including Captain John Sheridan. According to fan fiction, this drink was created at the Zeus bar at the Io Transfer Station in orbit around Jupiter.

• Brivari: Brivari is also a drink in the *Babylon 5* universe. It is a fine Centauri alcoholic beverage that requires precise climate control and seems to have similarities to wine on Earth. It is favored by Ambassador Londo Mollari.

• Ambrosia: This is an alcoholic beverage that appeared in the two Battlestar Galactica series. Ambrosia, which was green in color in the rebooted series, was prominently featured in that series in an episode entitled, "Tigh Me Up, Tigh Me Down," when Colonel Tigh's wife presents him with a bottle of the beverage and seduces him into getting drunk on duty.

Unlike some other aspects of science fiction, the fact that alcohol is so universal in this genre is not based on wild speculation of the future, but on well-established human nature that has not changed in thousands of years. As with sex, food, and even entertainment, science fiction authors do not even question whether space pioneers of the future will toast to their success or sit down at a bar after an exhausting day at work. Customs, technology, and society change dramatically over centuries, but many aspects of drinking and bar room culture remain similar. For example, if a resident of 1830s Boston were transported to the present, the time traveler from the early nineteenth century would most likely be completely perplexed and overwhelmed by virtually every element of society and technology. However, if they were guided to a typical pub-like bar in early twenty-first-century America, they would probably feel more comfortable—and have more of a connection between the age from which they came and the age to which they were transported.

One could argue that alcohol is only ubiquitous in science fiction and other literature because it provides a handy plot device. But even this is closely tied to the fact that alcohol—both the positive and destructive aspects—has been hardwired into human society for thousands of years and shows no sign of going out of fashion. And who knows, if we one day meet sentient aliens, maybe Gregory Benford's statement "Sentient creatures know what alcohol provides—a complex reaction that soothes your fears—and takes away the skittery anxiety that being intelligent means"[45] will be proven to be true for all such beings from whichever planet they hail.

# 3

# The History of
# Drinking in Space

Will science fiction authors' predictions of space saloons or alien drinks prove to be accurate? It is far too early to know the answer to that question, but alcohol has already had a long and complicated history in space exploration. This history has been influenced by public relations worries and safety concerns, and often by conflicting opinions among nations (including the United States, Russia, and other countries) regarding the appropriate role of alcohol in society and in space.

## *Prohibition in Space*

World space agencies, including NASA, currently prohibit alcohol consumption in space. There are varying reasons for this policy. Safety is certainly one of the biggest concerns. Astronauts have highly complex and extremely hazardous jobs, and the consumption of alcohol could impair their ability to perform critical duties safely. This, of course, is not unique to astronauts. For example, most professions that involve the handling of large and dangerous equipment—or responsibility for people's lives—have strict rules regarding consuming alcohol or certain medications while on duty or within a safe time period of performing their duties. For example, the Federal Aviation Administration has a strict eight hours "from bottle to throttle" rule. In other words, no pilot can fly an aircraft within eight hours of consuming alcohol. In some countries, this limit is as high as twelve hours—which is the same limit that American astronauts have to adhere to. Former Space Shuttle astronaut John Grunsfeld understands these concerns well, having flown on five space missions and performed some of the most complex and grueling space walks in history during repair missions to the Hubble Space Telescope. Grunsfeld has cautioned that "space is a very dangerous place. At any moment, it's important that you have complete clarity of mind and body to be able to respond to emergencies. You can also imagine the stresses of space and if someone ended up abusing alcohol, that could be a really terrible thing for being able to respond to an emergency. Certainly, for flying the Shuttle (or other spacecraft) you wouldn't want to have someone intoxicated."[1]

Public relations is also a significant factor in determining policies for alcohol in space. NASA is a high-profile government agency, and NASA and its astronauts are often highly scrutinized. Publicity often comes *not* when everything is proceeding perfectly but rather when something has gone wrong or when a scandal is perceived—real or imagined. NASA is therefore particularly concerned with maintaining the professional and disciplined image of its corps of astronauts. This relatively small group of individuals is highly accomplished in their respective fields, but they also symbolize an almost mythical status in American culture. As such, NASA is justifiably very protective of that image. For example, when reports materialized that two astronauts had allegedly been inebriated on the launch pad, it was a far more serious public relations challenge than would have occurred in any other agency. "That's a bad PR nightmare for NASA," commented former NASA astronaut Clayton Anderson. "They don't like to have to deal with stuff like that, just like diaper-clad ladies with pepper spray, and you'd expect with an elite corps of astronauts, you wouldn't have to deal with that sort of stuff. After all, we're all human and we all have to act responsibly."[2] In a political environment of stiff competition for budget dollars and constant challenges to keeping coalitions of support together, these types of issues can be a real risk.

Public relations and safety are not the only reasons that NASA restricts alcohol. In fact, the fumes from alcohol can actually be harmful to systems on spacecraft. Clayton Anderson explains that "NASA will tell you that alcoholic fumes in the environmental control system is not good—and that may be true. We don't use rubbing alcohol. We don't use hand sanitizer. We don't use anything alcohol based."[3] This policy was further articulated by Daniel G. Huot, a spokesperson for Johnson Spaceflight Center, during an interview he gave for the BBC. "Alcohol is not permitted onboard the International Space Station for consumption…. Use of alcohol and other volatile compounds are controlled on ISS due to impacts their compounds can have on the station's water recovery system."[4]

These types of restrictions are not only applied to alcoholic beverages. Since there are many sensitive systems on the ISS and other spacecraft, NASA is very careful about sending items into space that could inadvertently endanger the crew. Even astronaut care packages from home are highly controlled, with NASA being very particular as to what is allowed and what is not allowed to be delivered to astronauts in space. According to a NASA document addressing the rules relating to care packages, "Food is allowed. But, anything with significant crumbs.... Homemade treats like cookies—besides the crumb factor—can't go, because they're perishable and their quality can't be monitored. Products containing alcohol—not just alcohol to drink, but alcohol in perfume, aftershave, and mouthwash—can't go into space, and neither can cans under pressure—like shaving cream."[5]

NASA is in the business of taking risks, but it is also very careful not to take unnecessary risks—even ones that may seem trivial. If some of these restrictions seem excessive based on actual risk factors, it must be viewed in the context of both actual risk and public perception (which can turn into a policy and/or budgetary risk). In the end, even the appearance of scandal or unnecessary risk can potentially threaten funding from Congress and the Administration.

All these factors contribute to the prohibition of alcohol in space—and should not be dismissed. However, sometimes reality is far more complicated than what official rules and regulations seek to address.

## The Early Days of Space Exploration

While alcohol is currently restricted today, it was not expressly prohibited in space during the early days of the space program. In fact, alcoholic beverages flew into orbit during the Mercury Program, America's first human exploration missions in space, but these appear to have been intended as practical jokes rather than real efforts to provide libations to the first American space voyagers. Colin Burgess's book about the Mercury 7 mission, *Sigma 7: The Six Mercury Orbits of Walter M. Schirra, Jr.*, reports that Gordon Cooper and Indianapolis 500 racecar driver, Jim Rathmann, conducted a prank—or "gotcha"—on Wally Schirra during his Mercury mission. Prior to the flight, Cooper had "arranged to put a miniature, airplane-catering-size bottle of Cutty Sark scotch and one row of Tareyton cigarettes way in the back of a compartment of the instrument panel."[6]

As planned, Schirra discovered his surprise gift during his flight, but he did not consume the scotch in space. That said, he did not let the gift go to waste. "I drank the scotch as soon as I had a chance on board the recovery vessel,"[7] recalled Schirra. As a result, according to Schirra, the medics who examined him after his flight were surprised to find a tiny alcohol level in his post-mission blood work, but it is unclear whether they investigated this mystery any further.

## Apollo and Skylab

The booze smuggled onto Wally Schirra's mission may have been done surreptitiously, but a few years later alcohol was transported to space in a more official capacity, although it too appears to have been another astronaut "gotcha." Brandy was included with the Christmas meals for crew members Frank F. Borman, II, James A. Lovell, Jr., and William A. Anders on the Apollo 8 mission that orbited the Moon in December 1968.

The holiday meal for that mission featured turkey with gravy, cranberry sauce, and grape juice, as well as brandy that Deke Slayton had reportedly added to the meal containers. Specif-ically, three two-ounce bottles of 100 proof Coronet VSQ California grape brandy was stowed with the meals. The brandy was not destined to be consumed in space, however. Mission Commander Frank Borman would not allow the bot-tles to be opened. Even though Borman suspected it was intended to be a joke, he did not think it was funny. In a NASA oral history, Borman was quite clear about his feelings about the brandy, "When we opened up the dinner for Christmas and I found somebody had included brandy in there, you know, I didn't think that was funny at all. Because you and I both know, if we'd have drunk one drop of that damn brandy and the thing would have blown up on the way home, they'd have blamed the brandy on it. You know, I wanted to do the mis-sion and I didn't care about the other crap."[8] However, Lovell and Anders revealed later that they had no intention of consuming the brandy anyway, but apparently thought that Borman had overreacted to the gag.

The brandy returned to Earth and years later, in 2008, James Lovell auc-tioned off his bottle with Heritage Auc-

Small bottle of Coronet brandy that was included (but not consumed) in the astronauts' Christmas meal on Apollo 8 in 1968 (Heritage Auctions/ HA.com).

tion Galleries. The bottle was sold with documentation from Lovell that included a handwritten note that stated, "This bottle of brandy was included in my Christmas day dinner coming home from the Moon. Borman and Anders each had a bottle. To my knowledge, this was the only alcohol aboard Mercury, Gemini or Apollo spacecraft. James Lovell."[9] For this unique piece of space history, the bottle of brandy and accompanying documents were auctioned for $17,925. Based on the note that Lovell wrote for the auction, he apparently was unaware of the "gotcha" that had occurred during Wally Schirra's Mercury mission or an episode that would happen not long after his flight around the Moon.

Several months after Apollo 8, a bottle of alcohol did *not* remain unopened in space. One of the best documented cases (and possibly the first) of non-terrestrial drinking occurred during Apollo 11, the mission that landed humans on the surface of the Moon for the first time in history. When Neil Armstrong and Edwin "Buzz" Aldrin had successfully landed on the surface of the Moon, Aldrin wanted to perform a religious ritual to commemorate the historic achievement. Prior to the mission, Aldrin had asked the Reverend Dean Woodruff, the minister of the Webster Presbyterian Church (WPC) in Houston, to recommend an appropriate ceremony he could perform on the Moon. WPC had become known as the "Church of Astronauts" because of the large number of astronauts who worshipped there. In addition to Aldrin, additional astronauts who were members of that congregation included John Glenn, Roger Chaffee, Charlie Bassett, and others.

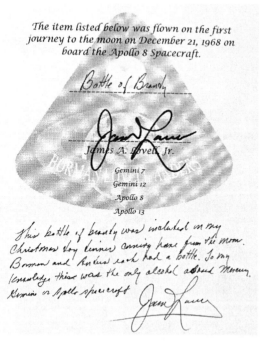

Aldrin and Woodruff decided that a communion ceremony would be most appropriate. Aldrin recalled, "I settled on a well-known

Note signed by astronaut James Lovell verifying that a bottle of Coronet brandy flew aboard the Apollo 8 mission in 1968 (Heritage Auctions/ HA.com).

expression of spirituality: celebrating the first Christian Communion on the Moon, much as Christopher Columbus and other explorers had done when they first landed in their 'new world.'"[10]

Prior to launch, Aldrin had informed Deke Slayton about his plans to conduct the ceremony. While Slayton had been the one who had stashed the brandy onboard Apollo 8 a few months earlier, he had concerns about generating too much publicity with this ceremony. He advised Aldrin to keep his public (broadcasted) comments from the Moon less controversial. Public perception was a major factor in Slayton's new attitude toward alcohol on U.S. missions. A few months earlier, there had been complaints when a Biblical passage from the Book of Genesis had been read by the crew of Apollo 8 as they orbited the Moon. As described in *The Huffington Post*, "He (Aldrin) originally wanted for the experience to be broadcast with the rest of his comments, but was discouraged by NASA which was at the time fighting a lawsuit brought by atheist activist Madalyn Murray O'Hair. She sued them over the public reading of Genesis by the crew of Apollo 8, citing the status of astronauts as government employees and the separation of

A communion kit (goblet and bag) that flew with Edwin "Buzz" Aldrin to the surface of the Moon during the Apollo 11 mission (courtesy David Frohman, president, Peachstate Historical Consulting, Inc.).

church and state to support her case."[11] As such, NASA was worried that if a Bible reading had generated such controversy, a full communion ceremony on the surface of the Moon with Bible readings *and* consumption of wine had the potential of making the Apollo 8 episode look insignificant by comparison.

Aldrin heeded the advice from Slayton and others at NASA and shortly after Neil Armstrong had declared, "The Eagle has landed," Aldrin made his public (on the air) comments by saying, "This is the LM pilot. I'd like to take this opportunity to ask every person listening in, whoever and wherever they may be, to pause for a moment and contemplate the events of the past few hours and to give thanks in his or her own way."[12] Once the broadcast had ended, Aldrin recalls, "I poured the wine into the chalice our church had given me. In the one-sixth gravity of the moon the wine curled slowly and gracefully up the side of the cup. It was interesting to think that the very first liquid ever poured on the moon, and the first food eaten there, were communion elements."[13] Aldrin then read Scripture (John 15:5), from three-by-five cards, "I am the vine, you are the branches. Whosoever abides in me will bring forth much fruit. Apart from me you can do nothing."

Regardless of NASA's hesitation to publicize the event, Aldrin's Apollo 11 communion marked the first—and probably only—time that an alcoholic beverage was consumed on a planetary body other than Earth. For his part, Aldrin did express some regrets years later regarding his decision to conduct a communion ceremony on the Moon. "Perhaps if I had it to do over again, I would not choose to celebrate communion." He added, "Although it was a deeply meaningful experience for me, it was a Christian sacrament, and we had come to the Moon in the name of all mankind—be they Christians, Jews, Muslims, animists, agnostics, or atheists. But at the time I could think of no better way to acknowledge the enormity of the Apollo 11 experience than by giving thanks to God."[14]

As with the auctioning of the Apollo 8 brandy bottle, the index cards that Aldrin read from were auctioned as well, and they sold for $180,000 in 2007. The chalice that Aldrin drank from, however, returned to the Webster Presbyterian Church. According to that church's website, members of that church in 1969 performed communion during the same hour that Buzz Aldrin had been conducting his ceremony on the Moon. Since that time, Webster Presbyterian Church has used the chalice each year to perform special communion ceremonies to mark the anniversary of Buzz Aldrin's lunar communion.

## Russia in Space

The United States does not have a monopoly on drinking in space, however. In fact, another country has attained a greater reputation for this practice. That distinction is claimed by the Russian Federation (and the former Soviet Union). What might surprise many Americans is that vodka is not the drink of choice for the Russians. Cognac became the preferred drink for cosmonauts—and they devised clever and elaborate methods for smuggling this contraband on board space missions. This became evident during the first reported alcohol consumption in space by the Russians that occurred in 1971 on the Salyut-7 Russian orbiting space station. As described in a 2017 article that appeared in *Russia Beyond*, "One of the cosmonaut's birthdays occurred while he was in orbit, and his friends gave a present before launch: they hid a bottle of Armenian cognac in the wristband used for measuring blood pressure."[15]

Alcohol was not the only contraband that would be elaborately smuggled on board Russian space missions. Whenever cosmonauts and support staff on the ground stashed "unapproved" items, they would make sure that it did not cause them to exceed weight limits for the missions or risk their spacecraft in any other manner. Some cosmonauts would go to extreme lengths to stay within weight restrictions but also assure that there was available liquor on board. According to Cosmonaut Igor Volk, some cosmonauts would go on strict diets to assure their preferred beverages could be smuggled onto the flight. "My partner Volodya Djanibekov and I thought of everything. A week before launch we didn't eat anything except bread and tea, and we lost almost two kilograms (4.4 pounds)."[16] As they were climbing into their space suits, they hid cellophane bags filled with two jars of pickled cucumbers and a bottle of cognac. Whether these extreme diets impacted their mission readiness is unknown, but this highlights how essential many cosmonauts considered cognac to be to a successful mission.

Fortunately, extreme weight loss was not always required to assure a drink in orbit. Cosmonauts would sometimes resort to methods that seemed to come right out of a spy novel. On some occasions, they would hollow out reading materials, such as thick book, and insert a bottle of about a liter and a half of alcohol. Care had to be taken when using this method because it could be easily discovered by even the most cursory investigation. If a book that contained alcohol was not picked up carefully, anyone nearby would hear it sloshing around—or "gurgling" inside. Volk recalled, "The most important thing is that it doesn't gurgle."[17]

The occurrences of alcohol smuggling were so frequent that subsequent crews would often find bottles hidden in space suits, behind panels, and in other locations. Cosmonaut Georgy Grechko recalled, "Once in a new physical education suit we found a half-liter flask with the label Eleutherococcus B (a natural tonic)."[18] In reality, to their pleasure, the flask actually contained cognac. "We calculated that each day before bed we could each drink 8.5 grams, but we managed to drink only half of the flask. We just weren't able to drink the rest."[19] This was not a result of lack of appetite. They had no choice but leaving half the flask unconsumed. The physics of space actually made this decision for them. According to Grechko, since the bottle and liquid were both weightless, after it had been halfway emptied, the cognac would turn into "froth" when they squeezed it and would not come out.

These smuggling practices continued for decades. According to private astronaut Richard Garriott, who flew to the ISS in 2008, "One of the historically common methods of taking a few extra personal items on board was on Soyuz, when you would be driving out to the launch pad, there is a point where—starting with Gagarin,—he stopped to unzip his space suit to urinate on the back tire of the bus. As it turns out, it was also an opportunity to push something inside your spacesuit at the last minute."[20] According to Garriott, this practice ended just prior to his launch, but there are still numerous other methods for contraband to be carried up to space. In fact, Garriott himself smuggled a bar magnet to space in the adult diaper he wore as he was getting suited up before launch.

As for the smuggled alcohol, Grechko revealed that the use of alcohol in space had been discussed at high levels within the Russian government. The Russian Ministry of Health had examined the merits of alcohol in space, and a senior representative from that Ministry stated, "In orbit, people have a very difficult emotional state. If before sleep, the guys drink 5–7 grams of cognac, I support it."[21]

In fact, many cosmonauts believe that cognac is *not* a danger in small amounts when consumed in space, and that it actually has many benefits. "The tiny drop of alcohol has a fantastic effect in space: it calms you down, removes the tension," recollected Valery Ryumin. "You quickly fall asleep and in the morning, you get up invigorated. This is better than a sedative, which you get addicted to quickly. The common opinion is that Armenian cognac is the best alcoholic beverage in space. I am convinced that it's necessary to legalize alcohol in space in small quantities, for example, as a sedative."[22] The need for sleep aids in space is well known. According to space medicine expert Saralyn Mark, astronauts "get an altered hormonal milieu—the internal background in the body where hormones work. Our hormonal profiles are based on a 24-hour circadian rhythm. As you go around the planet every 90 minutes, you've altered your sleep pattern. That may have an impact on your hormonal milieu."[23]

In contrast to the Russians, sedatives are often used by many American astronauts to get a full night's sleep in space. These sedatives can have similar or even greater troubling effects than alcohol. In fact, in the American space program there are a number of permissible substances that go up to space frequently with American astronauts. Specifically, sleep medications were regularly carried in the Space Shuttle pharmacy and are supplied on the ISS. "Sleep aids also cause impairment," commented John Grunsfeld. "Ambien is one of the better sleep aids, but is also well known to have odd effects with people doing things like sleep walking. One could argue that that's also a risk equivalent to alcohol."[24]

While cognac seems to be an unofficial staple of cosmonaut supplies, there does not appear to have been major abuses of this privilege. It has been utilized for an occasional sip to relieve stress or as part of social gatherings in space. For example, cosmonaut Valery Ryumin recalled that when a new crew came aboard Mir, protocol required that they hold a feast in space at which everyone would have a sip of cognac. Despite the more relaxed attitude of the Russians regarding drinks in space, Ryumin recalled that it still was not easy to stock these in-space social gatherings. "The most difficult thing was to take the bags to space, bypassing the various checks. Six liters were placed in certain caches; that's not a lot if you consider that the flight was to take a year and a half, and in this period there'd be another two missions that would bring six more cosmonauts to the station."[25]

## Sherry (Almost) on Skylab

As NASA completed the Apollo Moon landings and redirected its focus to other programs, it still did not have an official policy prohibiting alcohol consumption in space. For a short period of time, alcohol almost became an official part of NASA astronaut diets aboard the United States' new orbiting laboratory, Skylab. According to *The Astronaut's Cookbook*, Skylab was intended to be a "home away from home" for astronauts so that they could feel more at ease than on previous missions. To accomplish this goal, NASA food experts and nutritionists began the process of selecting appropriate food and beverages for the crew's food and drinks that would be an improvement over the rations that the Apollo astronauts had consumed on their much shorter missions. There were some people at NASA who did not believe the Apollo rations were fit for long-term human consumption. Don Arabian, who was NASA's spacecraft project manager, attempted to live on Apollo food for three days, after which he claimed that he had "lost the will to live."[26]

As the process of selecting more comforting foods proceeded, NASA food experts investigated whether wine would be a suitable beverage for crews, to make them feel more at home. The task of investigating potential wines for the Skylab menu fell to NASA food scientist, Dr. Charles Bourland. As explained by Bourland in his 2009 book, *The Astronaut's Cookbook*, "My boss was Mormon and consequently, the job of heading the wine selection process for the Skylab missions fell to me. Selecting a wine was an interesting project for the people in the food laboratory, and we had no shortage of volunteers for the taste panel."[27]

While many of the purported health benefits of wine were yet to be published, the NASA team believed that wine did in fact provide health benefits to the astronaut diet. Dr. Malcolm Smith, who was on Bourland's team, said "the question of whether wine promoted better health was still open." He also noted, "I would tend to believe that there is some value besides pure energy, either in the calming effect or promoting digestion. Somewhere in there, there's probably a beneficial effect from wine."[28]

Bourland and his team consulted various wine experts at the University of California at Davis, but this was not an ordinary wine consultation regarding the quality and characteristics of the wine. Perhaps the biggest challenge that Bourland and his team had to answer was whether they could find a wine that would remain unaltered when being subjected to the rigors of preparation and then launch into space. Any wine being flown to Skylab also required repackaging, which could impact its taste and overall quality. Additionally, the selected wine would need to endure the extreme shaking that occurs during launch.

With these requirements in place, sherry was selected as the best option. Sherry is less susceptible to the negative impact of repackaging as well as the substantial level of shaking and other extremes it would encounter before the crews would be able to consume it. According to a NASA oral history of Bourland, one of the reasons why sherry, a fortified wine, was more stable was because of the heating process it went though. As such, it also would not deteriorate after opening.

Thus, sherry seemed like the ideal official alcoholic beverage of the United States space program. Specifically, Paul Masson California Rare Cream Sherry was selected, and NASA had their initial supply of this beverage shipped to Johnson Space Center in Houston, Texas.

There was another challenge to overcome, however, that was unrelated to the quality and stability of the beverage. Gravity plays a large role on how

**Bottle of Paul Masson California Rare Cream Sherry that was originally intended to be part of Skylab astronauts' menus. The sherry never flew to Skylab (Charles Bourland).**

humans consume beverages (alcoholic or otherwise). In a weightless environment, astronauts are not able to consume any beverage from standard glasses as is done within the gravity of Earth. With no up or down in space, liquids do not pour out of bottles or settle in glasses as they would on the surface of our home planet. To overcome this challenge, NASA developed a special plastic pouch with a straw that would enable space-faring sherry drinkers to squeeze the container and drink their beverage. As will be noted in a later chapter, decades later this problem has still not been solved, and there have been numerous efforts to develop cups and glasses that can provide a more natural and pleasurable drinking experience in space.

The potential risks of consuming alcohol on Skylab were also a concern to the food specialists at NASA. They had great respect for their astronaut partners, but they also understood some of the obvious risks of providing alcohol on space missions. As such, NASA's food experts also wanted to assure that nobody over-consumed in space, so the astronauts would have been allowed only four ounces of sherry every four days.

The Skylab sherry project did not remain a secret for very long, as word got out about NASA's plan to send wine to space. A 1972 edition of *The Milwaukee Journal* wrote that "the era of prohibition is about to end in outer space,"[29] but Skylab astronaut Edward Gibson noted that they were also cautious about publicizing the prospect of sherry in space. "Let's just say that no one here is enthused about publicizing this thing any more than necessary.... The problem is that you have got some extremists around and we (astronauts) kind of represent a form of purity. As soon as you taint that purity with alcohol, they really get upset."[30]

As it turned out, Gibson had good reason to want to limit publicity. Ultimately, sherry never flew to Skylab because NASA became concerned about potential negative attention it might encounter if it were to become widely known that astronauts were drinking in space. In addition, when asked their opinions, most of the astronauts did not seem to care whether they had sherry on board. They felt that they would be too focused on their mission to spend a lot of leisure time sipping sherry.

Shortly thereafter, sherry was officially prohibited by NASA in two memoranda. An August 10, 1972, entry of a NASA document called "SP-4011 Skylab: A Chronology" states, "There was no basic requirement for including wine in the Skylab menu ... the beverage was not necessary for nourishment or to provide a balanced diet; it was not a fully developed menu item and would involve an unnecessary expense; it would aggravate a minor galley stowage problem ... and it would result in adverse criticism for the Skylab Program."[31] Although this document cites many reasons for the cancelation of the sherry, it is highly likely that the concern of "adverse criticism of the Skylab Program" was the most significant factor in this decision.

The document concluded that, "based on the above rationale, I am, by copy of this memorandum, notifying Deke Slayton that wine will not be included in the Skylab menu, and requesting [Dick Johnston] to immediately terminate all activity associated with developing and providing wine for Skylab."[32]

One other factor contributed to the ultimate decision that doomed the prospect of sherry or other wines on Skylab. NASA decided to test wine packaging on a KC-135 "Vomit Comet," a plane that simulates short periods of microgravity by flying parabolas in the sky. This packaging test did not go well. As stated in *The Astronaut's Cookbook,* "...the odors released by the wine, combined with the residual smell of people getting sick on the plane, had an unplanned effect on the crew."[33] Even many seasoned passengers on the "Vomit Comet" were reaching for their "barf bags." While this unpleasant reaction was not entirely the result of the sherry on the flight, it clearly was another "nail in the coffin" for the prospect of having alcohol on Skylab. When asked if he agreed with this decision, Charles Bourland stated, "No, but I had little choice. I had spent a lot of effort getting everything ready, I bought the sherry on a government purchase order and that required many signatures and an explanation from me for everyone. We had selected a product, developed a package for it and tested the package on the zero G aircraft."[34]

While the sherry never made it to Skylab, it did not go to waste. It was finally consumed by astronauts on the ground. In advance of the actual Skylab missions, the crews would participate in a simulated mission in a vacuum chamber. As told in their book, *Homesteading Space: The Skylab Story*, astronauts Owen Garriott and Joe Kerwin described the role the sherry played when they were participating in the SMEAT (Skylab Medical Experiment Altitude Test). "Fortunately for the SMEAT crew, however, by the time the decision was made to remove the sherry from the Skylab menus, the SMEAT menus had already been made out, and it was too late to go back through the process of completely rebalancing the various nutritional factors that would have to be changed if the sherry was removed. 'We had it,' (Robert) Crippen said, 'and we really looked forward to it.'"[35]

## No Beer on Mir—But Plenty of Cognac

Perhaps no other orbiting facility or spacecraft attained a more impressive reputation for cosmic drinks than the Russian space station Mir. While the Russian policy officially prohibited drinks on this facility, it was well known by many individuals associated with this program that alcoholic beverages—specifically cognac—were regularly being smuggled aboard. For many Russians, it was seen as an important element in crew relaxation and socialization.

According to cosmonaut Alexander Lazutkin, "During prolonged space missions, especially at the beginning of the space age, we had alcoholic drinks in the cosmonauts' rations." Russian doctors had recommended cognac to stimulate their "immune system and on the whole to keep our organisms in tone."[36]

As highlighted earlier, there were restrictions in place that still required imaginative smuggling methods to be devised, but Russian support personnel as well as Russian management were well aware that alcoholic beverages were being launched to Mir. Jeffery Manber, who today is CEO of Nanoracks, was one of the few Americans to work for the Russian "manned" space program and was a witness to this process. During the 1990s, Manber was working for Energia, the legendary company that is the prime contractor for the Russian human spaceflight program and, led by Sergei Korolev, was responsible for the Sputnik missions—including Yuri Gagarin's first human flight into space. Energia was also responsible for the Russian Mir space station program.

Manber was in Russia only a few years after the fall of the Soviet Union. It was in this timeframe, in the spirit of embracing capitalism, that Russia was attempting to commercialize Mir. In fact, in 1999, a company called Mir-Corp was created to help facilitate the commercialization of the Mir space station. One of their most notable achievements was to help enable the first paid tourist in space. American businessman Dennis Tito was originally scheduled to fly to Mir, but the Mir program was cancelled prior to his flight. Fortunately for Tito, he was able to get his flight into space, arranging to fly to the ISS through the American company Space Adventures in 2001. Tito paid an estimated $20 million for his voyage into space.

During Manber's time in Russia, the Shuttle-Mir program was also established. This program, which ran between 1994 and 1998, was a partnership between the United States and Russia in which the Space Shuttle would ferry astronauts to and from the Mir space station, allowing American astronauts to conduct long-term expeditions to that facility. As part of the partnership, American astronauts would also fly on Russian Soyuz spacecraft.

As it was the first partnership in space between the former cold war adversaries since the Apollo-Soyuz Test Project in 1975, it should not come as a surprise that cultural challenges quickly emerged. One of these cultural challenges was how each country regarded the role of alcohol in space. Manber recalls this well. "We instantly ran into some cultural differences. One of which is that the Russian and European side of the program—then principally men—they were up there 24/7 and come Saturday they have a day off and they would have some cognac or some vodka and watch some movies or read a book or just chill on a Saturday."[37]

Manber recalled that NASA was "absolutely aghast"[38] that alcoholic beverages were commonly present on Mir. While crew members would only have small amounts of cognac, no more than once a week after a hard day of work, this practice ran counter to NASA policy and philosophy on temperance in space.

It would be inaccurate to say that the Russians were alone in this relaxed attitude regarding consumption of alcohol in space. Manber recalled that NASA raised their concerns with the Russians *and* Europeans—particularly the French. Both countries share a far more "relaxed" view on the role of alcohol in society as well as in space.

As Manber explains, for the French, consumption of alcohol is an extension of their culture. "They have lovely cognac—I tried it. They worked very hard in the 90s to have high quality French food on the Mir space station. They brought in chefs who worked hard to make sure that the Foie Gras or even the basic food was good—and tasty."[39] The French did not want their astronauts—their national heroes—only eating the standard astronaut rations. "They wanted to treat them with French food and French cognac and other little delectables from France," explained Manber. "So, you saw the Mir space station as an extension of your country, of your cultures, and the French worked very hard at it—and they were very proud of some of the food they were able to bring up to Mir."[40]

In Manber's opinion, in the 1990s the Americans "brought their sensitivities and their cultural [feelings] to an international work environment which resulted in heated discussions on whether alcohol could be present on the Mir space station."[41] Manber saw this as conflicting with the spirit of what the international partnership on Mir was supposed to be.

This collaboration in space was certainly not the first time that cultural differences regarding alcohol have arisen between the United States and its partner nations. One only needs to observe the role of alcohol in the French Navy vs. the United States Navy. Like NASA, the United States Navy has a policy of Prohibition aboard all its ships. Unlike NASA, the exception to this rule in the United States Navy is when a ship is on a long deployment. According to Reuters, "if a vessel has been at sea for 45 consecutive days or more, sailors are allowed to have two beers, on a one-time basis."[42]

Compare this with the French Navy that has formalized the role of alcohol on their ships. For example, "There are no fewer than four bars on the *Charles de Gaulle* [a French aircraft carrier], where troops can purchase one alcoholic drink per day."[43] To assure that French sailors stay within their daily allowance, they developed an electronic tracking system. When asked why the French are so much more accepting of alcohol on Navy ships, French Navy Commander Lionel Delort replied, "We are French. Wine is very appreciated on board."[44]

Despite American conflicted attitudes regarding alcohol in society and in space, American astronauts on Mir were often willing participants in toasts or celebrations involving the obligatory cognac. On Space Shuttle mission STS-81 that docked with Mir in 1997, John Grunsfeld fondly remembers being invited to Mir for a social gathering. Grunsfeld recalled, "About mid-mission, while docked to the Mir space station, we were invited over to the Mir for a social event. When we were over there Valeri [Korzun] came up with a little bottle. And someone asked, 'Oh, is that vodka?' and Valeri said, 'No. no. We would never carry vodka to space. It's cognac.'"[45]

According to Grunsfeld, fine cognac came up to Mir in Progress supply ships in the medical kit. "It was fascinating because what he did was—these were tiny little bottles with no more than 250 milliliters. Just gave it a tiny squeeze and out came a singular ball, a round sphere, of cognac floating in the cabin. He gave each crew member a tiny little ball of cognac—no more than 25 milliliters each. It was a nice event and as far as I know, nobody got drunk at it."[46] Grunsfeld fully understands NASA's reluctance to accept any level of alcohol consumption on space missions, but also believes that moving forward—when humans are away from Earth for longer durations—these rules may have to be adjusted. "Properly managed, I don't think there is any real problem. Once you start living in space, I'm not sure there's any real hazard there over and above the tremendous hazard of space."[47]

Grunsfeld was not the only one who enjoyed these rituals aboard Mir. Michael Foale, a British (British/American) astronaut, visited Mir in 1997 and appears to have been perfectly comfortable imbibing occasionally. "Eating and drinking in space is more fun than on Earth. It is like camping: the social aspect of it is really important.... Even though the food is not great, it tastes good enough." Foale continued, "Same with drinking. If there is just a tiny little bit of alcohol on a station, boy, you enjoy that."[48] As always, the available drink was cognac. "You suck up a few drops with a straw and it spreads around the inside of your mouth and a bit up your nose because there is no gravity to pull it into your throat.... The alcohol gets absorbed quite instantly, like a cigarette draw where you get the instant buzz. Then, as it goes down your throat, you eventually get that warm feeling."[49]

The cognac did not always remain on Mir, however. In fact, on at least one occasion it traveled over to the Space Shuttle when it was docked at Mir. According to a NASA oral history about his time on the Mir space station, NASA astronaut Norman Thagard described an occasion when he and some of his fellow astronauts and cosmonauts had a drink on the Space Shuttle. Aware of the sensitivities of this topic, he told the interviewer, "I probably might not even ought to say this, but I will anyhow, because what can they do, sue us at this point?"[50] Thagard's fifty-second birthday occurred when he was on Mir at a time when the Space Shuttle happened to be docked. To commemorate the event, the Russians gave him a "really good Russian cognac."[51]

The cognac had originally been purchased for a party that had taken place before the launch in Russia, but a large quantity of cognac remained unconsumed. Thagard and some of his Russian colleagues "decanted" the remaining cognac into plastic containers and wrapped the containers tightly with tape. According to Thagard, "So we labeled them all 'juice,' and they carried them out and put them in the Soyuz. So, we launched with quite a lot of cognac on the Soyuz."[52]

When the Shuttle had docked, Thagard explained that he and his fellow astronaut Charles Precourt, as well as cosmonauts Vladimir Dezhurov (Veloga) and Gennady Strekalov, all went to the Spacelab module that was in the cargo bay of the Space Shuttle. However, Thagard thought that they could not all disappear to Spacelab without notifying mission commander, Hoot Gibson. Thagard approached Gibson and said "Hoot, I just want to tell you that..." Apparently Gibson anticipated the question. Gibson replied with a laugh and said, "I don't want to know about it."[53]

Thagard commented, "I guess, and between the four of us we had a fifth of cognac, but I never saw anybody drunk. In fact, I never saw anybody drink more than what I would consider to be one shot.... We all had our straws, and we would take straws and we would stick them down into one of the bubbles and just take a sip, and then you would gently push the bottle and it would float over to the next person, and they'd put their straw in."[54]

In June 1985, a bottle of wine made a historic voyage into space aboard the Space Shuttle *Discovery* (STS-51-G). Specifically, a half-bottle (375mL) of 1975 Château Lynch-Bages accompanied French astronaut Patrick Baudry as part of the first human spaceflight collaboration between France and the United States. As such, the French wanted their culture to be represented in space. This included some French modifications to the usual astronaut menu. According to *Decanter* magazine, the "French influence was ringing out loud and clear in Baudry's choices—jugged hare, langoustine, crab mousse, Cantal cheese and chocolate mousse."[55] In a 1985 interview, Baudry stated, "Our food tradition is very ancient.... It is part of our civilization. It is very important. In America you don't have that same concept of food."[56]

As for the bottle of wine, it almost was prevented from flying. "But not because of NASA," commented Jean-Michel Cazes, the owner of Château Lynch Bages. "It was the French government who initially disapproved. They wanted to concentrate on showing France's technological and engineering prowess instead of wine and perfume."[57]

Baudry was quite passionate about ensuring that wine was part of the mission. His motivation was heightened even more when he learned that Coca Cola would be flying into space soon. Baudry stated, "I thought it would be a shame if it (Coke) arrived up there before wine, which represents our culture, our savoir-faire, and even a part of the history of humanity. So I did everything in my power to take some wine with me, in a symbolic way, and I succeeded. We didn't make the most of it, unfortunately, as you can't drink any alcohol on American spaceships. I'm sure it would have been a different story with the Russians (laughs)."[58]

In addition to the bottle, a vine leaf and ten small vials of the 1983 vintage of Château Lynch Bages were given to the crew, but since drinking alcohol in space was prohibited, none of the wine was consumed. However, the bottle of Château Lynch Bages was probably the most traveled bottle of wine in history, having orbited the Earth 111 times and traveling 2.9 million miles.

## Fire and Collision on Mir

As John Grunsfeld cautioned, astronauts need to be ready at a moment's notice to have complete "clarity of mind."[59] One such moment occurred on February 24, 1997, when fire broke out on Mir. According to the *New York Times*, "The fire burned like a blowtorch for about 14 minutes and blocked the exit leading to one of two lifeboats, each capable of holding just three people."[60] Nobody was seriously injured during this episode, but after the fire was out, astronaut Jerry Linenger recommended everyone drink powdered milk as well as vitamins, which apparently could serve as a countermeasure to any toxic materials that the crew may have ingested during the fire. The Russians also received some medicinal advice from their ground controller who suggested to take "a little special medicine,"[61] which was apparently a non-subtle code for cognac. It is believed that the fire was caused by a faulty igniter on the solid-fuel, oxygen-generating canister and caused one of the most dangerous episodes in the history of human spaceflight.

By all accounts, Mir was a highly stressful environment and the fire was the start of a series of near disastrous occurrences, including a collision with a docking cargo ship a few months later. It is doubtful whether the existence of alcohol on that orbiting facility was in any way related to these incidents, but what is clear is that when it was noted in the press that cognac was on Mir, various news stories seemed to imply that it was part of the problem and questioned the sobriety of the Russian cosmonauts. Again, perception rather than verifiable fact often appears to have been the real danger of having alcoholic beverages on board. Despite this bad press, what is clear is that cognac played an important ceremonial role in helping to create bonds between crew members from the United States and Russia. This contribution should not be dismissed since it came at such a critical time in space exploration and relations between the two rival super powers. It served as an essential precursor to the partnership that currently exists at the International Space Station. One certainly cannot conclude that providing liquor in space is always a good policy, but it has served an important role in international missions.

## The International Space Station

As the United States and its international partners started assembling the International Space Station (ISS), they wanted to project the perception and reality that the ISS was a model of efficiency and international cooperation. While there were many technical, political, and budgetary challenges as the Space Station was assembled, the orbiting laboratory has largely achieved this goal and has become a model of what the international community can accomplish through partnerships in space.

Unlike on Mir, NASA wanted to remove the perception that alcohol had any place on this new facility. By most accounts, smuggling and consumption of alcohol on the ISS has been far less prevalent than it was on Mir, but this does not mean that the ISS remained dry for very long. American astronaut Clayton Anderson, who lived on the ISS for five months in 2007, explained that "NASA will tell you that there is no alcohol aboard ISS. As a person who lived there for five months, I'll tell you that's bogus."[62] As was the tradition on Mir, cognac remained the preferred space drink. "We would celebrate big days of success with just a couple sips of cognac during our evening meal. We never abused the privilege."[63] Private astronaut Richard Garriott, who flew to the ISS in 2008, expressed similar opinions on this topic. "I personally experienced alcohol in orbit, but it's important to point out that any alcohol that I am personally aware of is a tiny, tiny bit in contrast to the number of days on orbit. No one is abusing alcohol up there. No one is becoming an alcoholic up there. Nobody is getting drunk up there."[64] Garriott explained that there was a small amount of both vodka and cognac on the ISS when he was there and when his crew arrived on the ISS, they received a toast to welcome them. They would also occasionally have sips during meals. "Is it required? Of course not. But I do believe that just like on Earth the web of human experience includes alcohol and when people are living in some cases years at a time in space, I think a prohibition on that is unnecessary and ill-advised"[65]

While Clayton Anderson understood the reasons why NASA was not anxious to publicize this fact, he was also agitated by the level of "denial" that would occur with some of his superiors. He cited the tendency of some individuals to say "I don't want to see it. I don't want to hear about it. I don't want to talk about it"[66] to establish plausible deniability.

Despite these occasional exceptions, Jeffrey Manber of Nanoracks expressed similar disappointment with NASA's attitude at the ISS as he did with their policy at Mir. "To this day, I'm a bit disappointed at how little we project the ISS for what it is—humanity's outpost in space. Which means it's filled with human beings and human brains—and they bring their cultural values and to this day there is a little bit of a divide. The Europeans and the Russians have a much more relaxed view on habits while relaxing."[67]

Moving forward, as plans are made to send humans back to the Moon and on to Mars, the social and stress-relieving elements that small amounts

of alcohol have played on Mir and elsewhere should not be dismissed out-
right. As John Grunsfeld noted, if it is managed well, it might be appropriate
to have a certain amount of alcohol on future long-term missions—like the
Navy policy on long deployments.

Clayton Anderson echoes this sentiment, noting that "if you have a large
settlement of people living on the Moon or on Mars or even on a space station,
personal creature comforts are important. The guys on (the Space) Station—
they watch TV and movies and they're getting first run movies sent to them.
Things that remind you of home, that give you the sense that you're not so
far away, are tremendously important."[68]

## Pre-Flight Traditions

Irrespective of the role of alcohol in space, drinks have always played a
part in the rituals of astronauts and cosmonauts on the ground. In the early
days of the American space program, the astronaut corps consisted of test
pilots who were known to blow off steam with drinks. This was quite under-
standable given the fact that they risked their lives practically every day. Some
of the most memorable—and real-life—scenes in the film and novel *The Right
Stuff* took place at Pancho's Happy Bottom Riding Club (or Rancho Oro Verde
Fly-Inn Dude Ranch) at Muroc Air Force Base (now Edwards Airforce Base)
that was run by the colorful early female aviator, Florence Lowe 'Pancho'
Barnes. She once reportedly said, "Flying makes me feel like a sex maniac in
a whorehouse with a stack of $20 bills."[69] While she did little flying at this
time in her life, her establishment was famous for serving up copious amounts
of drinks to innumerable test pilots.

The level of ritual drinking clearly diminished as the American astronaut
corps moved away from the test pilot culture, but certain traditions that
included alcohol remained. Perhaps the most established tradition is a gathering that is held at the "Beach House" on the grounds of the Kennedy Space
Center at Cape Canaveral, Florida, a couple days prior to a launch. The ocean
front house was purchased by NASA in 1963 (along with some other homes)
when the launch facility was expanding. Unlike the other homes, this house
was saved from demolition and was converted to serve as a location for astronauts to socialize and relax. The Beach House (also known as the Kennedy
Space Center Conference Center) serves as the location of a prelaunch tradition for crews. According to John Grunsfeld, "On the U.S. side, in the beach
house a couple nights before a launch we have our spouses and families for
a reception. There's usually a bottle of champagne or wine involved that the

crew signs and it's then displayed in a cabinet. So, at the beach house there is this historical record of crews through wine that they toast with before a launch."[70]

The Beach House and the prelaunch reception should in no way be construed as a place where astronauts go to get drunk. It is a quiet and isolated location where astronauts can mentally prepare for their flights. According to astronaut Mike Mullane, "This is sacred sand out here, it really is. It's where people have made their final goodbyes, and they were final. But we know it. I mean, there's nobody, no spouse, no astronaut walks that sand that doesn't know there is a possibility that this is forever.... That's the reality of an astronaut's life, and spouse's life, in those final days and hours before a mission. Fear and joy overwhelm you."[71] And, heartbreakingly for the families of the *Columbia* and *Challenger* crews, this would be their final goodbyes.

Space Shuttle astronaut Scott Kelly also mentioned there were other prelaunch rituals—not involving alcohol. "Before a Shuttle launch we used to play a low-ball poker game where the commander of the crew had to lose a hand before you could walk out of the suit-up room. The idea being that if he lost, he'd given up your bad luck for the day."[72]

A cabinet at the NASA Beach House at the Kennedy Space Center filled with bottles signed by astronaut crew members days before their space missions (Christopher Shaffer).

The Russians have also had prelaunch drinking traditions for decades. John Grunsfeld, who trained in Russia and visited the Mir space station, recalled, "On the Russian side, there are various toasts prior to the launch that all the cosmonauts and management do before space."[73] Fellow astronaut Scott Kelly recalled some of these toasts, traditions, and rituals in his 2017 book *Endurance*, which told the story of his nearly one-year stay on the International Space Station. Since the Space Shuttle had been retired in 2011, Kelly had to be transported to the ISS on a Russian Soyuz spacecraft in 2015. In his book he recalls his 1,600-mile flight from the cosmonaut training center in Star City (45 miles outside Moscow) to the Baikonur Cosmodrome in Kazakhstan, where the Russian space agency launch their cosmonauts into space. This flight that carries the crew, support personnel, and VIPs has a tradition of partying all the way to Kazakhstan. According to Kelly, "Everyone has been drinking since we left Star City this morning.... Russians never drink without eating, so in addition to vodka and cognac they are serving tomatoes, cheese, sausage, pickled cucumbers, dried salty fish chips, and pieces of salted pork fat called *Salo*."[74] He recalled that on his first flight to Baikonur in 2000, when he was making a trip to the bathroom, he was stopped and compelled to drink shots of a Russian moonshine called samogon.

Many of the traditions are extremely regimented and started early in the space program, often directly related to what Yuri Gagarin did prior to his historic launch. This is highlighted in a 2014 *BBC* portrait of the early Russian space program that described cosmonauts Pavel Belyayav and Alexey Leonov being awakened for their launch of Voskhod-2 on March 18, 1965. "They were given a medical and then observed several traditions that had developed in the years since Gagarin's launch."[75] As part of this tradition, Yuri Gagarin himself opened a bottle of champagne, filling glasses of the cosmonauts on hand and they all took a sip. Like the American tradition, they all signed the bottle, but in the Russian tradition, they recorked the bottle and pledged to finish the bottle when they returned safely to Earth. During their orbits, Alexey Leonov would perform the first space walk in human history.

Many of these extensive traditions have continued to the present day. Scott Kelly thinks these types of traditions are important. "There is a fine line between ritual and superstition, and in a life-threatening business such as spaceflight, superstition can be comforting even to the nonbeliever."[76] In addition to taking a sip of champagne the morning of the launch, they continued the rituals started by Gagarin. Cosmonauts also sign the doors of their rooms at the Cosmonaut Hotel on the day of launch. Gagarin's bodily functions are also sacred among cosmonauts. Apparently, Yuri Gagarin had to "take a leak" on the way to the launch pad. As a result—and to honor this unintended tradition—all male crew members now get off the bus and urinate on one of the tires. Kelly describes his experience with this ritual. "I center myself in front of the right rear tire and reach into my Sokel suit. I really don't have to pee, but it's tradition…. The tradition is so well respected that women space travelers bring a bottle of urine or water to splash on the tire…."[77]

These are not the only traditions on the way to and at the launch site. Richard Garriott recalls, "You get on the bus to the launch site and you have to watch *White Sands in the Desert*, an old spaghetti western supposedly Gagarin watched…. When you get to the launch pad, you're now at the base of a fully fueled rocket for the first time—in the US, no one is allowed at the launch site except the closeout crew…. In Russia, not so. There is a small brass band there, Russian generals, my father was there. As you begin to climb up the gantry, another one is that you get a kick in the seat of the pants… The guy who closes the hatch is literally the same person that closed the hatch for every cosmonaut starting with Gagarin—he closed my hatch too … he was mostly toothless and quite old."[78]

## NASA Drunk Astronaut Report

While occasional sips of cognac in space do not appear to have impacted crew performance or caused any level of inebriation, alleged ground-based inebriation has caused more embarrassment for NASA. Abuses will occasionally occur in any profession, but NASA has good reason to strive for an astronaut corps that exemplifies responsibility and honor. However, in 2007, NASA was challenged with a potential scandal that could have tarnished that image. NASA was rocked by rumors and the findings of a report that alleged that there may have been occurrences of irresponsible consumption of alcohol in the days leading up to the launch of the Space Shuttle. According to a report of the NASA Astronaut Health Care System Review Committee, "Interviews with both flight surgeons and astronauts identified some episodes of heavy use of alcohol by astronauts in the immediate preflight period, which has led to flight safety concerns. Alcohol is freely used in crew quarters. Two specific instances were described where astronauts had been so intoxicated prior to flight that flight surgeons and/or fellow astronauts raised concerns to local on-scene leadership regarding flight safety. However, the individuals were still permitted to fly."[79] The report also highlighted that the current process of issuing medical certification of astronauts is not structured to detect irresponsible consumption of alcohol prior to launch. The report recommended that NASA create better policies that would make individuals and their supervisors accountable for irresponsible alcohol consumption. It also recommended that there should be an alcohol-free time prior to launch and that a mechanism should be put into place to monitor and appraise any concerns raised by key individuals.

While NASA accepted many of the recommendations from this report, they also disputed whether astronauts were ever intoxicated on launch day. During a NASA press briefing on August 29, 2007, Bryan O'Connor, NASA Chief Safety and Mission Assurance Officer, outlined his investigation of these claims. O'Connor reviewed over 1,500 anonymous reports, looked at tens of thousands of mishap and close call records, and spoke to dozens and dozens of relevant people in all phases of an astronaut preparation for launch. After this review, he stated, "I was not able to verify any case where an astronaut spaceflight crew member was impaired on launch day or any case where any NASA manager disregarded recommendations by either a flight surgeon or another crew member that an astronaut crew member not be able to fly in the shuttle or the Soyuz."[80]

Russia was also implicated in this report. As such, and since one of the astronauts cited was said to have been inebriated prior to a Soyuz launch, the Russians also took issue with the report. Igor Panarin, from Roscosmos, the Russian Space Agency, told the Associated Press, "We categorically deny the possibility that this could have happened at Baikonur.... In the days at Baikonur before the launch, this is absolutely impossible. They are constantly watched by medics and psychiatrists."[81] Although cosmonauts do sip champagne as part of a pre-flight tradition, it is not an amount to cause any level of inebriation. In fact, the highly rigid pre-flight rituals that the Russians perform would seem to add a compelling level of validity to the Russian denials.

That said, the Russians did not use this incident to deny that they are more indulgent regarding consuming alcohol in space. In a 2007 interview in *The Telegraph*, a Russian spokesperson said, "Folks are flying in orbit for half a year or so. They do tons of work. Especially exhausting are space walks where they can lose several kilos in several hours. That's why, many of us think that it won't be that bad to have a little drink to replenish one's strength."[82]

It is unclear if any astronauts were intoxicated on the launch pad, but this report came shortly after NASA was rocked by another scandal involving Astronaut Lisa Nowak, who had allegedly driven 900 miles in a diaper to attack a romantic competitor for the affections of another astronaut. This occurrence does not appear to be alcohol related, but it was yet another highly publicized black eye for the preferred heroic astronaut persona.

## Conclusion

There is a common thread that runs through all the stories cited of drinking in space, whether imbibed by Americans, Russians, or others. In all the known occurrences, astronauts and cosmonauts have consumed alcohol responsibly and in small quantities. This is not to say that this will always be the case—or that there have not been occurrences that have not yet come to light—but the current data shows that our space-traveling professionals have lived up to their reputations of acting extremely professionally in space. In fact, many of the instances of sipping cognac in space have served an important bonding role among crews of different cultures—and have taken the edge off the stress of a highly challenging and dangerous job. Nonetheless, moving forward, other factors will influence the rules and availability of fermented beverages in space.

As humans venture farther and farther away from Earth and privately orchestrated missions and facilities become more commonplace, the consumption of alcohol in space may well be determined by the marketplace. However, what is clear by looking at over fifty years of human spaceflight is that human explorers will always find a way to imbibe some quantities of alcohol in space, regardless of official rules and policies.

### Polar Bars

#### Club 90 South

It may not be the Restaurant at the End of the Universe, but Club 90 South is one of the most remote drinking establishments on Earth and is probably the best analog in existence of what an early Mars saloon may look and feel like. This unusual drinking hole resides at the Amundsen Scott South Pole Station in Antarctica. According to an article in *The Atlantic,* the bar is "a simple, wood-paneled joint with a hole in the wall opening up to the outside, where the bartenders would put the Jagermeister to keep it chilled. Massive pallets of beer, wine, and liquor were flown in with the winter crew, and they prayed it would last ... them all nine months."[83]

To be clear, Club 90 South is not the only pub in Antarctica. In fact, the much larger McMurdo Station boasts three bars to provide libations to the inhabitants of this base. During the prime research months, McMurdo

**Patrons of the Club 90 South at the Amundsen Scott South Pole Station in Antarctica (Phillip Broughton).**

boasts a population of over 1,000 researchers and support staff. The Amundsen Scott South Pole Station is much smaller, with a maximum population of roughly 150, much closer in size to potential early settlements on Mars. As such, the Club provides an interesting case study of both the possible positive and negative aspects of early drinking establishments on the Moon or Mars.

Phil Broughton, a health physicist from California, participated in a field season in Antarctica in (2003) and inadvertently became a part-time bartender at Club 90 South when he sat behind the bar and ended up serving patrons drinks. According to Broughton, "You see things that leave you uncomfortable. There were a good dozen people who were drinking to kill the days—that was hard to watch, and it was hard to serve. Though at some level, I'd rather have you drinking in front of me than drinking on your own."[84]

Heavy drinking is far from a universal problem at these Antarctic outposts, but it is a problem that has long been a concern during stays in Antarctica and could very well become a concern on early space settlements. One possible way to minimize this problem is to eliminate one of its main causes—boredom. Planetary scientist Chris McKay noted that boredom is the primary reason for overdrinking in these environments. "From my personal experience in many field trips to many extreme environments, the primary reason folks turn to drink is that they don't have meaningful work to focus on…. Keep everyone busy doing meaningful work, and no one gets bored and drinks to excess."[85]

### Haughton Mars Project

By contrast, the Haughton Mars Project (HMP) on Devon Island in the Canadian Arctic has taken a different stance on alcoholic consumption. HMP has restricted alcohol consumption during its activities. According to its official alcohol policy, "The HMP is a 'dry' project. No alcoholic beverages may be brought to, dispensed at, and/or consumed at, HMP, including 'privately.' This policy is strictly enforced. Alcohol consumption results in impairment in senses and judgment, in mood changes, and in reduced physiological tolerance to environmental stress (to cold temperatures in particular), all of which result in dramatic increases in risk to the safety of all field participants, let alone to the productivity, comfort, and well-being of all."[86]

Alcohol was once permitted during off-hours at this site, but it was prohibited after some alcohol-related safety concerns occurred according to HMP Principal Investigator Pascal Lee. However, the HMP field activities only occur over the summer each year. As such, the Antarctic model is probably a more realistic model for long-term stays or settlement at such locations as Mars.

# 4

# Space Beer

## Introduction

As described in the earlier chapter, despite restrictions on drinking alcohol in space, astronauts and cosmonauts do occasionally imbibe adult beverages during their space missions. However, there is a big difference between the sporadic smuggling of small amounts of booze onto launch vehicles and the prospect of actually manufacturing alcoholic beverages in space. Manufacturing any significant amount of alcohol away from Earth will require resources that do not currently exist in space. As such, space-based brewing and distillation would seem to fall within the realm of science fiction, but this assumption is not entirely correct. Real efforts have been underway for decades that may enable future space-based fermentation, brewing, and distillation.

Manufacturers face many challenges if they hope to successfully manufacture alcohol away from the safety (and gravity well) of Earth—particularly on far-away locations such as Mars. Science fiction author and futurist David Brin believes that some of the biggest challenges facing space booze manufacturers include "purity of raw materials—'The Martian' would have needed a LOT more water to flush out his Martian soil; then growing the ingredients. 'The Martian' could make vodka from potatoes, but little else. Then, ensuring the yeast strains suit conditions like gravity, will bubbles of $CO_2$ be able to escape the mash, in lower-G? Then purification, filtration and, if you like, distillation. All will be different, calling for ingenuity, luck … and 'sciencing the crap out of it.'"[1] Nonetheless, while Jeffrey Kluger from *Time Magazine* is also skeptical about large-scale manufacturing of alcohol anytime soon, he thinks small-scale production is highly likely, noting, "If there are two dozen people on Mars, and they've got yeast and they've got grain, sugar, and water, it is not that hard for scientifically-minded people to manufacture alcohol."[2]

The carbonation problem mentioned by Brin was put to the test in the 1980s when the "Cola Wars" spilled over into space. In 1984, the Coca Cola

Corporation negotiated an agreement with NASA to send a specially designed Coke can into space that would allow astronauts to consume carbonated beverages in microgravity. Upon learning of this venture, however, Pepsi did not want to be left behind on Earth. Pepsi approached NASA to assure that they also were able to test a cola dispenser in space. According to *Mental Floss*, Coca Cola developed a 12-ounce can with a special nozzle and valve on top of the can at a cost of $250,000. Pepsi invested quite a bit more, spending $14 million to design and manufacture their can, which also had a valve on top to dispense the carbonated contents.

Both cola experiments were launched in July 1985 on the Space Shuttle *Challenger* (STS-51-F). Unfortunately, the astronauts on board this mission apparently were not very impressed with the results. According to a University of Southern California article, "The astronauts thought both cans of soda were horrible. The sodas were warm due to lack of refrigeration and if the astronaut burped, some of the soda came up as well."[3] While some of these complicating factors were completely out of the control of the manufacturers, this was not a tremendous start for those seeking carbonated beverages in space.

Coke would go on to fly its products a few more times, including a trip to the Russian Mir space station in 1991 and a test of a redesigned dispenser on the Space Shuttle *Endeavor* in 1996. Regardless, carbonated beverages are still not served in space decades later. In fact, private astronaut Richard Garriott had to contend with this restriction during his 2008 flight to the International Space Station (ISS). A student had asked him what it would be like to open a soda can in space. "I thought that would be a very interesting thing to see,"[4] says Garriott. However, he was not allowed to bring a soda can up so instead he brought an Alka Seltzer. According to Garriott, "I was able to see the foamy water and the bubbles staying in the middle of it."[5]

Manufacturers will also need to contend with the fact that many astronauts report that their taste buds change in space. This dilemma is clearly not as critical as life support and other life-dependent systems, but appreciation of food will play an important factor in the psychological health of crew members on long space missions. During a 2016 panel discussion hosted by the space advocacy non-profit, Explore Mars, Inc., in Washington, D.C., noted geomicrobiologist Penelope Boston highlighted the importance of food and taste in exploration and in isolated environments. "Comfort food takes on an immense significance—I've done a lot of expeditions—we're dealing with grim backpack meals. I always have my [butter nuts with me]. I'm not a very food-oriented person, but it takes on a significance far beyond what I experience in my daily life, because when you're in a very challenging environment, part of your psychology needs to revisit [hominess]. I think it's some-thing common to all of us."[6]

What makes this issue even more complicated is that there is no con-sensus regarding the extent of the impact of the space environment on taste; it seems to impact each person differently. When asked about this issue, Vickie Kloeris, NASA's Food Manager for the ISS said, "That depends on who you talk to. There is no scientific evidence that microgravity alters the taste of food. There is anecdotal evidence from crew members that they feel like their taste buds are somewhat dulled in orbit. Other crew members say it's all in their head and there is no difference."[7] Astronaut Clayton Anderson did not notice much of a change in his sense of taste. The changes he did notice he believed were attributable to the "fluid shift to his head and the stuffy nose and the fact that you can't smell as well."[8] Fellow astronaut and physician Scott Parazynski commented, "It's the same as having a cold or allergies ... a stuffy nose definitely dampens your sense of smell and conse-quently your sense of taste."[9] Richard Garriott has his own theory. "If you have a hot plate of food [on Earth], those chemicals are rising from the food directly toward our nose. So putting our nose over the item going into our mouth, we get a chance to smell it more acutely. In space, as you know, those same molecules will be distributing in all directions, as opposed to upward with warmth. It makes me at least ponder that one of the reasons why people need a boost to their flavor sense, is because of how molecules disperse ... that full participation of what we think of as the eating process?"[10]

Regardless of the cause, many astronauts compensate for the perceived reduction in the sense of taste by heavy use of strong spices and sauces in their food; spicy meal enhancers appear to be quite important to the ISS crews. According to Garriott, Sriracha Hot Chili Sauce was quite popular on the ISS. Astronaut Peggy Whitson jokingly "threatened to bar entry to the crew of the visiting shuttle *Atlantis* unless they came bearing a promised resupply of the spicy stuff. Only when shuttle commander Jeff Ashby announced that he had the goods did Whitson say, 'Okay, we'll let you in then.'"[11] While this was intended as a joke, it did highlight how important such food enhancers are to the residents of the ISS.

Despite these issues, and since Prohibition still "officially" reigns in space, one would think that endeavors focused on alcohol production in that environment would also be prohibited as well, but this is not the case. Numer-ous companies (large and small), academic institutions, and private ventures around the world are now investigating how alcoholic beverages could be produced in microgravity as well as on planetary bodies such as the Moon and Mars. While some of these projects are clearly intended as stunts to help publicize a specific brand, others are being conducted for the genuine purpose of investigating whether the space environment can benefit their Earth-based product—or if one day it will be possible to ferment adult beverages entirely off-world.

It can be argued that the seeds of the current interest in "space booze" were first planted decades ago, in the 1970s, when word got out that NASA was considering adding alcohol to the Skylab menu. This news was of immediate interest to some alcohol producers. According to former NASA food expert Charles Bourland, some alcohol producers approached NASA, hoping to associate their products with space exploration. One winemaker sent Bourland a high-end case of their wine, but since sherry had already been selected, the wine could not be used. Not wanting to waste such a fine case of wine, they covertly deposited it in the office of an astronaut who they knew was a wine aficionado and had been an advocate for having wine in space. He would have to settle for consuming this wine on Earth, however.

Bourland also recalled that a representative from a well-known alcoholic beverage producer visited him at the Johnson Space Center in Houston requesting to leave a sample, hoping that their product would be launched to Skylab. Bourland told him that "there was no possibility that it would ever be considered, but he (the representative) insisted on leaving it, so I took it and placed it in a cabinet. A few days later, I got a call from my NASA boss."[12] Apparently, the liquor company had released a press release stating that their product was being considered for space. According to Bourland, "My boss was upset and demanded that I have the sales rep come by and pick up the bottle and he did and apologized for getting me in trouble."[13]

These early attempts by booze manufacturers to have their products sent into space were almost certainly for publicity reasons. It would be many more years before an alcohol company was part of an actual space mission to investigate the science and process of how alcohol might be manufactured in space and how alcohol reacts to the space environment.

## The First Beer Yeast in Space

Beer has played a prominent role in human society for thousands of years. As such, it is fitting that beer was the focus of one of the first space-based alcohol-related experiments. Perhaps it should also not come as a surprise that this space beer experiment was led by a college student. In the early 1990s, a graduate student at the University of Colorado named Kirsten Sterrett firmly placed herself in the history of space booze when she began to ponder how yeast would perform in the microgravity of space. "It was mostly a result of circumstances of the time. I graduated with a bachelor's in aerospace engineering in 1992 and there were not a lot of aerospace jobs available at the time."[14] As a result, Sterrett ended up getting a temporary job as a fermentation technician in the yeast lab at Coors. "I think I was an unusual applicant for a temporary lab technician. Everyone laughs about the rocket scientist. The fact that I even applied for that position was a bit odd. They probably should have passed me by for someone far more qualified."[15] As it turned out, she became quite interested in her temporary work with beer yeast and started brainstorming. "I was very excited about how the beer was created and fermented and I thought that would be an interesting experiment to do in a microgravity environment."[16] Thus, her graduate thesis topic had been created. However, deciding on her topic was not the biggest challenge she had to overcome. Her biggest challenge was how to get her beer experiment flown in space.

Fortunately, Sterrett was not new to space-based experiments. As an undergraduate, she had worked for a company named Bioserve Space Technologies that had been established at the University of Colorado in 1987 to develop hardware and research related to life science in microgravity. At Bioserve, Sterrett had worked on an experiment to investigate how alfalfa grew in the microgravity environment.

As she had developed the concept of yeast in space at her temporary job, she asked her Coors colleagues if they might be willing to supporting her concept. "I asked the people in the yeast lab whether they might be interested in helping with my graduate research and might be interested in beer in space."[17] It did not take too much convincing on her part. Her Coors lab colleagues agreed and helped support her unique beer yeast experiment.

Sterrett and her team would not be starting from a blank slate, however. This would not be the first time that yeast had been sent into space; but to their knowledge it would be the first time yeast was sent to space as part of a beer-related experiment. The first yeast experiment was launched along with cosmonaut Gherman Titov, with the Soviet Union's second manned space mission in 1961, Vostok-2, but this experiment had been compromised. Although some changes were observed in the yeast samples, Sterrett noted in a research paper, "It was determined after the flight that the observed changes in yeast viabilities were a result of storage conditions … and not to the effects of space."[18]

The Soviets were not deterred by this initial failure. They sent yeast into space again on Cosmos 368 in 1970, this time with better results. "It was found that space flight did not significantly alter radiation effects on yeast, that yeasts from spaceflight could recover from radiation doses similar to the way ground control samples did...."[19]

The United States entered into the space-yeast arena a couple of years later on Apollo 16 in 1972. On this flight, an experiment was included that looked at yeast genetics and morphology. Over the following years, several other yeast-related experiments were launched into space by the United States and the Soviet Union (later Russia), but Sterrett's experiment was the first that was specifically intended to understand the viability of future beer production in space. In addition to her partnership with Coors, she also partnered with her old employer, Bioserve. They were essential in designing the experiment and assuring that the experiment made it to space. Sterrett recalled, "I was very lucky that I was a 19- or 20-year-old already doing work in a microgravity environment on the Shuttle. It was an amazing opportunity for a young engineer."[20]

Bioserve must have liked Sterrett's unusual proposal because the beer yeast experiment was launched on two separate Space Shuttle launches in 1994. In February 1994, eight fermentations of lager yeast were launched on STS-60, and then in March 1994 four more were sent up on STS-62.

Several varieties of commercially available lager yeast flew in both liquid and dry forms. While on board the Space Shuttle missions, the liquid samples were fermented in Bioserve's Fluid Processing Apparatus (FPA), designed to conduct fluid tests in microgravity. On these missions, the FPA was modified to allow the release of the carbon dioxide that would be created during the fermentations process.

When the samples returned to Earth, Sterrett said, "We repitched [repitching is recovering yeast cake after fermentation and re-culturing for additional fermentations] it for the beer and measured the beer's specific gravity and checked how the yeast performed.... And we did a total cell measurement—the viability of the cells, how many of the cells were still alive. And we did some protein tests to see if the beer expressed any other types of proteins that they didn't on the ground samples."[21]

After their initial repitch, the samples from the first shuttle mission did see "significant reduction in the suspended cell growth rate, decreased peak cell count, and decreased rate of suspended cell sedimentation."[22] After the fifth repitch, the yeast had returned to its ordinary state.

In the end, there was not any significant difference between the flown yeast and the base sample that remained on Earth. "I think everyone who does an experiment wants to see night and day differences and be able to say 'Aha! This is the one thing that I can totally prove!' I was hoping to see something life-changing but didn't. We did find a stress protein expressed, but that makes sense because when you're in an environment you're not used to, you're going to show some stress."[23]

Studies like this had (and still have) real value. According to Louis Stodieck, who is now director of BioServe, "What we're trying to do now is to find the specific mechanism of that (increased fermentation efficiency in space), and then we can ask whether we can modify the fermentation process on Earth to take advantage of that—or is it possible that we could genetically engineer an organism to mimic what it does in space?"[24]

At the end of the experiment, Sterrett's curiosity prevailed, and she tasted the space samples. According to her recollections, "I always said I wouldn't do an experiment that I couldn't eat or drink in the end." It was barely enough to taste—1mL—but, "after the experiment was all done, I gave (the space-beer) a little taste ... but why throw something like that away?"[25] Apparently, the space beer did not taste very good, however.

Even though beer had been the catalyst of this experiment, Bioserve went on to send other yeast experiments into space. According to Stodieck, "We know from subsequent space experiments sponsored by Bristol-Myers Squibb Pharmaceutical Research Institute that the efficiency of producing fermentation products increases [in a weightless environment], in fact quite significantly."[26]

Today, more than twenty years after the beer yeast experiment was conducted, the greatest impact of this experiment may not have been the science that was generated or any greater understanding of manufacturing beer in space. The true legacy may have been the numerous space-alcohol related projects that would follow years later. When asked about her thoughts on the numerous projects that may have been inspired by her project, Sterrett said, "I think it's awesome. When people research, they invent, and new things are created. I think that's where the human race is always supposed to go. Always learning. Always researching. Always doing new things."[27] If she were given the opportunity to conduct the experiment today, she replied, "I think larger scale experiments. Mine were on the order of millimeter fermentations. Going to bigger volumes would be very interesting."[28]

## Japanese Space Beer

After the experiments conducted by Kristen Sterrett and her partners, more than ten years passed before any substantial advancement was made in the pursuit of space booze. In 2006, Japanese beer company Sapporo became an early corporate entrant into the space beer business. Unlike Coors, who did not aggressively publicize their audacious support of a space beer experiment, Supporo fully understood the publicity value of space beer. Rather than yeast, however, Sapporo opted to send barley, one of the most common crops for beer production, into space. While this would seem to be an unusual project for a beer company, Sapporo prides itself on a history of adventure and innovation.

Japan's oldest beer company, the Kaitakushi Brewery, was founded in 1876 (renamed the Sapporo Beer Factory in 1886) by Seibei Nakagawa, who was Japan's first German-trained brewmaster. Nakagawa left Japan at the age of 17, in an age when it was rare for someone to leave that country in pursuit of wealth, adventure, or anything else. Eventually, he made his way to Germany and became a trained brewmaster. With this knowledge, he returned home to Japan; becoming the only German-trained brewmaster in Japan. To this day, Sapporo embraces Nakagawa's spirit of adventure in their advertisements. As stated on the Sapporo website, "The legend of our beer began with the adventurous spirit of Seibei Nakagawa."[29]

With this adventurous backstory, it seems appropriate that Sapporo would be interested in pushing into new beer frontiers with their space barley experiment. The Japanese beer giant partnered with Okayama University and the Russian Academy of Sciences, who were running a study examining how to grow edible plants in microgravity aboard the International Space Station (ISS). Crops examined in this study included wheat, peas, and lettuce. Barley was also added to this list, but the primary reason for its selection was not focused on beer production. It was chosen because it can grow in diverse environments and temperatures. Nonetheless, barley in space was still of paramount interest to Sapporo, whose product is dependent on barley crops, and in the end, Nijo malting barley, a variety that was developed by Sapporo, was selected for flight.

On April 24, 2006, a total of 0.9 ounces of barley samples was launched to the ISS aboard the Russian Progress-56 supply ship. Once on the ISS, they resided on the Zvezda Service Module until September of that year. After several months in space, a portion of the seeds was placed in the root module of the LADA greenhouse chamber (see Chapter 6), where they were irrigated with water and given a daily light/dark cycle of 20 hours of light and 4 hours of dark. After 26 days of growth, the barley had thrived, reaching a height of 50–60 cm, which was very similar growth to the ground-based control experiment that was being conducted simultaneously.

The flown barley samples were returned to Earth on September 29, 2006, aboard the Soyuz TMA-S and began to be analyzed. According to a research paper written about the experiment, the results suggested that "the barley germinated and grown in Lada onboard ISS is [was] not damaged by [the] space environment, especially oxidative stress, which is suspected to be induced by space radiation and microgravity."[30]

From the Sapporo perspective, they also determined that there was no apparent difference in taste or genetic makeup between the space barley and regular barley.

In April 2007, Sapporo seeded the first generation of space barley (0.14 ounces) at a research farm northeast of Tokyo and for the next few years continued planting subsequent generations of barley, the descendants of the original "space barley." In 2009 they harvested the fourth generation of their unique crop, and in September of that year they held tastings of barley tea at locations around Japan.

Only a few months later, they released the limited edition of a beer that had been brewed using this barley. The December 3, 2009, press release stated, "Sapporo Breweries Ltd. (Tokyo, Japan) launched the sale of the world's first beer produced using malt made 100% from 'space barley,' the progenies of spaceflight barley seeds. This limited offer was exclusive to internet sales and residents of Japan and proceeds went to charity. Only two hundred and fifty customers were selected from a lottery to purchase the new 'SAPPORO Space Barley.'"[31]

Despite the novelty of the space barley beer, Sapporo also wanted to assure that a top-quality beer was produced. According to Sapporo's Director of Strategy, Junichi Ichikawa, "There's really no beer like it because it uses 100 percent barley. Our top seller is the Black Label brand, using additional ingredients such as rice. This one doesn't, and is really a special beer."[32]

## *The King of Beer—on Mars*

> "While Budweiser can currently be enjoyed in every corner of the world, it is time for the King of Beers to set its sights on the next destination—Mars. We believe that life in space deserves to be filled with the same enjoyments we have down here on Earth—including beer."
>
> —Ricardo Marques, Budweiser[33]

In Gregory Benford's 1988 short story, "All the Beer on Mars," early Mars explorers brew their own beer using a portion of their supplies and some smuggled yeast. Eventually, they toast their discovery of life on Mars with their interplanetary home brew. When humans are in reality traveling to Mars and beginning initial settlements, will home brewing be a realistic option if these settlers would like to indulge in a cold beer? If a major American producer has its way, significant beer production will take place on the surface of Mars.

At the South by Southwest festival on March 11, 2017, in Austin, Texas, American beer behemoth Budweiser announced their aspirations to become the first beer manufacturer on Mars. The announcement came during a moderated panel by actress Kate Mara (*The Martian*). The panel featured Anheuser-Busch's vice president of Innovation, Valerie Toothman, retired Astronaut Clayton "Clay" Anderson, and Budweiser vice president Ricardo Marques.

**Mission patch of the third Budweiser barley experiment launched to the ISS in December 2018 (Anheuser-Busch).**

According to the Budweiser press release based on this event, "39 million miles away, the next generation of space exploration awaits its first pioneers. And while life on the Red Planet is still in the near—yet distant—future, Budweiser just unveiled its own ambitious commitment, to create microgravity beer for when we make it to Mars. Known for raising a cold one to the American Dream and those who work hard for it, Budweiser wants to be part of this monumental journey in reaching the next frontier."[34]

Budweiser's ambitions were not limited to a press conference, however. In December of 2017, they joined the growing number of companies sending alcohol-related experiments into space. Budweiser partnered with a Kentucky startup called Space Tango and the Center for Advancement of Science in Space (CASIS) to arrange the launch of their barley-based research project.

The experiment, called the Germination of ABI Voyager Barley Seeds in Microgravity project, was designed to analyze the impact of microgravity on the growth of barley; one of the essential ingredients of beer. On December 15, 2017, the experiment began its voyage into space with CRX-13 (commercial resupply service mission) launched aboard a SpaceX Falcon 9 rocket to the International Space Station. The barley experiment was contained in Cube-Labs provided by Space Tango. (Similar to CubeSat, a Cube Lab is a 10 cm × 10 cm × 10 cm container that "can house a single experiment, whether it's from the life sciences, material sciences, or another field of study.")

According to Dr. Gary Hanning, Director of Global Barley Research for Anheuser-Busch, "Barley is a key raw material in Budweiser and frankly in most beers as well, so it made natural sense to begin our research there. Mastering how barley grows in a different environment is a crucial challenge for us in bringing Budweiser to Mars ... experiments were carefully designed with our partners to focus on seed exposure and seed germination, to evaluate how barley, a key ingredient in the iconic Budweiser recipe, reacts in a microgravity environment."[35]

One of the CubeLabs contained 3,500 dry barley seeds to be analyzed upon their return to Earth. The other CubeLabs contained test tubes designed to allow barley seeds to grow while on the ISS. The experiment lasted for 30 days to look at the effects of microgravity on the seeds as well as on germination and seedling phases. Space Tango Program Manager Gentry Barnett stated, "This partnership will not only produce scientific data that could lead to barley production improvements on Earth but could also lead to the first beer produced on Mars."[36]

While results from their experiments have not been released, Gary Hanning revealed that the space environment may have impacted the barley: "While we cannot officially confirm, it appears that the seeds exposed to the micro-gravity environment do not grow as expected."[37]

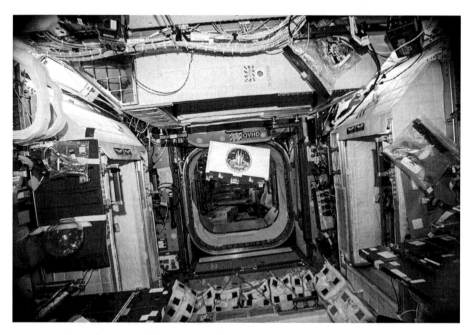

"Bud on Mars" barley experiment floating in the microgravity of the International Space Station (Anheuser-Busch).

Whatever Hanning meant by "do not grow as expected," Budweiser faces many challenges if they hope to produce (and consume) beer in microgravity and on Mars. As noted earlier, one major problem in microgravity is how carbonated beverages react in a weightless environment. Without the pull of gravity, the gas in carbonated drinks does not react the same way as it does on Earth. In an interview with *Inverse,* Vickie Kloeris, who is subsystem manager for the ISS food systems at Johnson Space Center, commented, "The problem with beer and with any other carbonated beverage in microgravity is how to keep the carbonation evenly distributed throughout the beverage.... The tendency in microgravity will be for the carbonation to separate from the beverage itself."[38] This can literally result in a "foamy mess." Budweiser hopes to "conduct studies on how beer and carbonation will likely behave on Mars," which has only one-third the gravitational pull of Earth. To put this challenge to the test, they will conduct "experimentation with ingredients in space, with the eventual goal of creating a microgravity version of Budweiser suitable for brewing and drinking on Mars."[39] It remains to be seen, however, how the beer giant will be able to utilize microgravity to design a brew specifically for Mars gravity.

As mentioned previously, astronauts report changes in the sensitivity of their taste buds. Budweiser also plans to address this issue and is taking steps to assure consistent quality even in space. According to Hanning, "Preserving our signature Budweiser taste is a top priority. The taste of Budweiser has not changed in the past 141 years—something we take a lot of pride in. We're committed to creating a beer that tastes just the same in space, as it does on Earth."[40]

The beer giant clearly hopes to be the first to provide beer to future space settlements, but they also have near-term goals for this research. As they investigate the impact of the space environment on barley and potential other elements of beer production, their aspirations also come back to Earth. "We're constantly researching different strains of barley here on Earth, as well as the different climates our seeds are exposed to,"[41] stated Gary Hanning. "We see microgravity as a learning opportunity as we work to cultivate new strains of barley that can thrive across the globe. Our research will offer insights on steps to creating beer on the Red Planet, and also provide valuable information on the production of barley and the larger agricultural community here on Earth."[42]

Budweiser's space ambitions did not end with the 2017 launch. Budweiser sent a third experiment to the ISS in December 2018 (the 2017 launch had two experiments). According to an ISS U.S. National Laboratory statement, "Budweiser will send its third experiment to the ISS National Lab, this time examining strains of barley seeds as they go through the processes of steeping (rapid hydration), germination, and kilning (drying) associated with malting barley."[43] Budweiser hopes to develop malt barley variants that can grow in more extreme environments, which could be beneficial for future Mars crops as well as here on Earth. As for future goals, Hanning stated only, "We are excited to continue working on the 'Bud on Mars' project and look forward to what's next."[44]

## Why Beer on Mars?

> "It's hard to imagine a beer manufacturer NOT wanting to be the first to develop a bona fide 'zero-gravity beer' that has less carbonation—so to avoid any 'wet burp' catastrophe in the spaceship!"
>
> —Ian O'Neill[45]

Why would a traditional beer manufacturer decide to initiate this project? The answer to this question may have to do (at least in part) with recent trends in beer consumption. According to a Beer Institute poll, beer sales have declined over the past few years, particularly among millennials. According to the poll, drinkers currently choose beer 49.7 percent of the time opposed to the mid–1990s when drinkers chose beer 60.8 percent of the time.

As beer sales decline amongst millennials, the "Bud on Mars" project could be an effort by Budweiser to capture the imaginations of the younger demographic. It may help Budweiser to appear more "hip" and attractive to younger people. Well-known tech billionaires such as Elon Musk and Jeff Bezos have become symbols of innovative thinking—of challenging old established ways of doing business, of being the biggest "disrupters"—in recent years. Budweiser may hope to capture some of that magic to energize the company and bolster the next generation of beer consumers. "Companies like that always want to be at the forefront of their technology. They want to be recognized as innovators; new idea types of people," commented former NASA astronaut Clayton Anderson. "Again, it goes back to that theme of 'you want to bring a little bit of home with you,' so kudos to them for thinking outside the box."[46] This opinion is echoed by Budweiser Vice President Ricardo Marques, who stated that by pursuing this project, Budweiser is celebrating the entrepreneurial spirit that helped found Anheuser-Busch back in the nineteenth century.

Perhaps the biggest question for Budweiser and other beer companies interested in projects of this kind is whether this project (or similar ones) will help to improve beer sales. A recent Rasmussen poll suggests that it probably will not. The poll found that only eleven percent of American adults say they are more likely to buy a beer that is brewed on Mars. By contrast, thirty-three percent of individuals polled stated that they would be less likely to buy a space beer. Forty-five percent said that origin would have no impact on their beer consumption decisions. The precise language of the poll question was "Would you drink beer brewed on Mars?" The accuracy of this poll is impossible to guess. The real test is when Budweiser or other companies are able to provide tangible results that they can share with their customers through their products. Budweiser may well be playing the long-game, hoping for a substantial advantage in the future.

Sales are always a part of any corporate strategy, but one must not discount the prospect that Budweiser is genuinely passionate about being the first supplier of beer on Mars and having a real opportunity to play a role in the next step in human expansion. According to Gary Hanning, "We believe the American spirit to pursue new interplanetary destinations is alive and well and when people get to Mars, we want Budweiser to be there. Budweiser has perfected our liquid here on Earth and we can't wait to see what the future of beer looks like, especially as we consider the future of space tourism and the role beer will play in bringing people together—in space and beyond."[47]

If Budweiser is really committing to a long-term strategy with the expectation that humanity will settle Mars in the next several decades, they are positioning themselves to play a dominant role in off-world beer production and sales if that market one day becomes significant. As Marques concluded, "When the dream of colonizing Mars becomes a reality, Budweiser will be there to toast the next great step for mankind."[48]

## Vostok Space Beer

Budweiser has some competition in becoming the "king of space beer," however. This rival to the throne resides in Australia, but Mars is not their goal. Their focus is squarely creating a beer that is specially made for a microgravity environment. In 2011, an Australian aerospace company began an unlikely partnership with an Australian brewer to create a new type of beer that became known as Vostok.

The idea of the space beer pre-dated Vostok by several years. Jason Held, founder of Saber Astronautics, was an engineer based in Colorado and was working on the Wide Field Camera-3 of the Hubble Space Telescope. As is common in Colorado, he and many of his colleagues were home brewers.

"We used to sit around and talk about what it would be like to have a beer in space."[49] Years later, in 2009, Held was living in Australia and had just founded Saber Astronautics when he went for a drink at a local pub at Manly Beach, near Sydney, Australia. While there, he met the founder of the newly formed 4 Pines Brewing Company, Jaron Mitchell, and they instantly hit it off. Held asked him, "I'm going to ask you a question that I've never asked a man before … would you like to have your beer in space?"[50]

As luck would have it, Held had found a kindred spirit. Mitchell had grown up wanting to pursue a career in space exploration, but there were no opportunities in West Australia to fulfill this goal. Both decided that they wanted to find a way to combine their love of space and beer.

After brainstorming various options, they agreed to attempt to brew the first beer that was *specifically* designed to be consumed in space. To do so, they had to ferment a product that would overcome the noted challenges that are unique to conditions of microgravity, specifically the fact that carbonated drinks do not react the same way in space as they do in Earth's gravity and that astronauts often report changes in their taste palates.

This combination would not make for a pleasant drinking experience for weightless beer consumers. To address this challenge, the final product would need to have a reduction in carbonation, allowing customers to consume their product in low gravity conditions, without carbon dioxide–induced discomfort. However, they would still need to retain enough carbonation so that consumers could feel it on their tongues, "so that it feels like a beer, not an alcoholic tea,"[51] commented Held. This would be a tricky balance to achieve. As testing proceeded, they observed that when the carbonation level in beer was changed, it also impacted the flavor of the beer. Thus, finding the right balance was not just a matter of selecting a beer with a stronger taste and reducing the carbonation level.

Eventually, they selected a beer that was firmly in "The Goldilocks Zone"—it seemed to be "just right." That beer was 4 Pines Dry Irish Stout, which is described as "a dry Irish style stout, presenting almost black & bearing a generous tan head. Aromas of coffee, chocolate and caramel are matched with a full-bodied mouthfeel, a smooth finish and rounding bitterness."[52] Since the name "4 Pines Dry Irish Stout" did not inspire visions of space exploration, they decided that a new name was needed. They called it Vostok, after the first human spaceflight in history, Vostok-1, that launched Cosmonaut Yuri Gagarin. Thus, Vostok Beer was founded in 2010.

With the beer now selected, they needed to find a realistic method to test it. According to Held, "I did a literature review of food consumables and alcohol. There was a really big gap in the literature for alcohol consumption.

Maybe our bodies will metabolize alcohol in space at a different rate."[53] While it is well known that cosmonauts and astronauts have imbibed alcohol in space, none of these occurrences were "official," thus no research had been conducted to monitor them as they drank. The best research that he found was conducted by the Federal Aviation Administration in the late 1960s and early 1970s to examine how people metabolize alcohol differently at high altitude. They had three groups of test subjects who consumed alcohol (and a placebo) at ground level, at 12,000 feet, and at 20,000 feet. Each was given either a non-alcoholic simulant, a moderate level of alcohol, or a larger level. As a result, they found some differences between alcohol absorption effects between the subjects at high altitude and at ground level when they consumed the higher level of alcohol.

Held and his team decided to "take the same principles of this experiment and apply it to an on-orbit type of experiment. What is the alcohol absorption rate in microgravity? We don't know."[54] As a result of the alcohol restriction in space, they would not be able to test their product on the International Space Station. According to Held, "NASA is right to be conservative, since alcohol can be abused, and the body's limits in space are not known. So, this research is not just to make a beer to enjoy, but also to learn how people's drinking limits change in microgravity. Knowing these limits is the only responsible way to allow explorers to drink under any condition. Because if anyone deserves a good beer, it's them."[55]

At the time, the burgeoning space tourism business was generating a lot of excitement. Originally, they were hoping to test their beverage with a company like Virgin Galactic or some other commercial company, but those services were not yet available. Without access to space, their only other option for weightless conditions was on a zero-G flight through Zero Gravity Corporation (ZERO-G).

Based on the famous "Vomit Comet" that NASA trained astronauts on, their Boeing 727 creates moments of weightlessness by performing parabolas starting at an altitude of 24,000 feet. "The pilot then begins to pull up, gradually increasing the angle of the aircraft to about 45° to the horizon reaching an altitude of 32,000 feet. During this pull-up, passengers will feel the pull of 1.8 Gs. Next the plane is 'pushed over' to create the zero-gravity segment of the parabola. For the next 20–30 seconds everything in the plane is weightless."[56]

Finding their weightless test environment was only part of the challenge. Assessing their product would not be as simple as having one of their employees drink a few beers on a parabolic flight. They needed to create a plan that was controlled and as scientifically valid as possible. Otherwise, it would be considered nothing more than a stunt.

Even though they were operating on an extremely tight budget, they needed to hire an individual who had experience in zero-gravity flights to conduct the experiments. Held explained, "It was called the vomit comet for a very good reason. We needed to find someone who had very good experience in parabolic flight and we needed someone who can drink booze and knows the difference between microgravity discomfort and alcohol discomfort. Not someone who is a lush, but someone who has had good casual drinking experiences."[57]

Coincidentally, a new company had been founded a few months earlier called Astronauts4Hire (also known as A4H; now the Association of Spaceflight Professionals). Inspired by the numerous burgeoning commercial spaceflight companies entering the scene, the goal of this group was to start "a new collaborative effort to support and develop a pool of qualified, commercial scientist astronauts...."[58] A4H had about fifteen people on their roster at the time, but they had not yet found their first customer. Held was particularly anxious to hire them not only because they provided the experience he needed, but also because he wanted the Vostok project to be the first client to hire this new and unique organization. According to Brian Shiro of Astronauts4Hire, "This project put us on the map. We were a new organization and they said, 'Hey, we want to hire you for this job right away!'"[59] In addition to providing the first pool of test candidates, A4H also helped them set up the test protocols to assure that the projects would be able to maximize their time on each zero-gravity flight.

Eventually A4H and Vostok narrowed their list of beer drinking test subjects down to two people—a primary "subject" and a backup "subject"— and they were ready to conduct their first flight.

Before any of this could happen, another significant challenge had to be overcome. Perhaps the most challenging obstacle was not creating the beer or hiring the crew or even performing the mission. The biggest obstacle may have been government and bureaucratic "red tape." The team had to address a tremendous amount of regulatory challenges before they could fly this experiment—particularly since this would be a biomedical test on a human subject. All biomedical experiments with human test subjects are subject to meticulous government oversight. Eventually, the FAA did assign someone to oversee the experiment, because they wanted to verify that there was real research being done and not just "fun in the sky."[60]

With regulatory issues accounted for, the "subject" selected, and their product ready for flight, the Vostok experiment was just about ready for launch. As part of final preparations, a week before the flight they performed a baseline test with their private astronaut (the subject) to observe how his body reacted in normal conditions. They provided him with six beer samples, each with 150mL of beer, and observed how his body metabolized it. They also had him fast the night before the flight and *not* take any anti-nausea medications that could skew the results.

Despite these preparations, once the flight began, they realized that there were some problems with their overall process. "What we found was that there were no problems drinking the amounts he was drinking in microgravity," recalled Held. "We did find that the experiment setup made him more uncomfortable than it should have been. It wasn't an easy flight."[61]

Crew members and passengers of previous parabolic flights have noted that microgravity discomfort does not typically occur during the weightless portion of the flight, but during the high-g portion when passengers are experiencing the Coriolis effects. According to the *Merriam-Webster* online dictionary, Coriolis force is defined as "an apparent force that as a result of the Earth's rotation deflects moving objects—such as projectiles and air currents—to the right in the northern hemisphere and to the left in the southern hemisphere." In standard zero-gravity flights, this effect is often mitigated when crew members and passengers recline on the floor of the aircraft during the high-G periods of the maneuver. However, because he was the only person tasked with performing the experiment, the test subject was *not* able lay down on the floor during the high-G times. He used these periods to prepare the experiment for the next period of weightlessness. As a result, "around the third sample, he was feeling a bit of discomfort, but it wasn't from the alcohol, but from the microgravity discomfort."[62] In fact, according to Jason Held, after resting for a little while, the test subject drank another sample and was feeling well again—the beer had settled his stomach. But this did not last long. At the top of the final parabola, he experienced some negative Gs and hit the ceiling, which caused all the beer to be ejected from his stomach.

Despite these difficulties, they gained some valuable knowledge. In fact, they observed a difference between the baseline experiment and the flight experiment but decided not to publicize this because they only had one data set.

The Vostok team also gained knowledge on how to improve the experiment during future tests in order to prevent much of the "non-alcohol related" discomfort that had occurred in test number one. In a follow-up flight, they would allow the subject to have a light breakfast as well as a half dose of the anti-nausea medication. They also decided to allow two people to participate on the second flight, allowing the test subject to rest during the high-G portions of the flight while the other crew member prepared the beer doses.

Another change was made for the second flight that took place in 2013. This time, the crew members were selected internally. Biochemical engineer Tara Croft would be the test subject and Jason Held served as the flight scientist. With the new protocols in place, the second flight went much more smoothly, but Held concedes that they still did not have enough data to release any definitive report. They will need to conduct many more flights to make any conclusions about alcohol absorption in microgravity. "From a medical perspective these are case studies, enough to say 'something is likely different,' informing a hypothesis. Alcohol absorption rates also vary based on ethnicity, age, sex, etc … but changing the gravity does seem to be having some effect."[63] Nonetheless, there did appear to be a difference between the absorption in microgravity and Earth gravity consumption.

Regardless of whether they had solid data regarding alcohol absorption in microgravity, they released Vostok beer well before the second flight took place. Public sales began in Australia in May 2011, generating solid sales, and the beer even won first prize in the 2012 Australian International Beer Awards.

The beer was not the only product to emerge from this project. During the first flight, the "subject" had considerable problems getting the beer out of the bottle that he was supposed to be drinking from because of the weightless conditions. As a result, the Vostok team designed a special insert that was eventually used for a new microgravity bottle. According to Held, "The

Vostok team testing beer bottle inserts on a Zero G flight in 2013. Tara Croft and Jason Held are pictured in the foreground (Vostok Pty Ltd.).

vision is to be able to pull a bottle off the shelf and bring it on to your space plane or whatever vehicle and not have to do anything else."[64] To accomplish this goal, they had to contend with engineering and physics challenges such as fluid flow and surface tension, as well as the NASA requirements for space-flight consumables. After years of design, in 2013 they successful tested a couple of prototypes on their second zero-gravity flight; one bottle that had been 3-D printed and one bottle that had been constructed out of stainless steel.

In 2018, Vostok launched an Indiegogo campaign to raise $1 million to fund pre-sales of their space beer bottle. According to their description on the crowdfunding site, "Having successfully created a beer that can be drunk in space, we now need to make the bottle so astronauts and future space tourists can drink in space just as we do on land."[65] Unfortunately, the Indiegogo campaign only made a little more than $31,000. This was far short of their goal, but they were able to generate a significant level of press and excitement, receiving a readership of over 500 million.

Vostok has taken some valuable steps to understand how well humans metabolize alcohol in space, as well as creating more user-friendly containers to allow the astronauts to drink from.

Jason Hand said, "Since there are physical issues with drinking beer in space (carbonation, flavor, etc.), we realized that there was an opportunity not just to support the space tourism industry, but fellow beer lovers as well.... The thought of making the world's first space beer is really exciting for us— it's a real contribution to making people happy for a long long time."[66] Held and the Vostok team believe that alcohol is inevitable if we start to settle other worlds—and not just beer. "Vodka certainly makes sense as a space alcohol—with a few tweaks it's usable as a rocket fuel, so there's some fun thought experiments there. But at the end of the day, people on Earth love beer, therefore people in space will also want beer."[67]

Vostok plans to move forward with their ambitious goals and is planning to send three crew members on their third zero-gravity flight in 2019.

## Ground Control Beer

Not to be outdone by Aussie brewers, the Ninkasi Brewing Company in Eugene, Oregon, was inspired with the prospect of brewing beer using ingredients (or descendants of ingredients) that had flown in space.

Founded in Springfield, Oregon in 2006 by Jamie Floyd and Nikos Ridge, they named their new beer company the Ninkasi [nin-kah-see] Brewing

Company after the ancient Sumerian goddess of fermentation. Floyd and Ridge were able to quickly grow the business and their reputation, then moved the operation to Eugene, Oregon, in 2007, where they continued to grow. In fact, the Brewers Association ranked Ninkasi the 38th largest craft brewer in the United States in 2017.

Building a successful brewing company might seem sufficiently challenging for most young beer producers, but Ridge and Floyd had much higher (literally) goals with the creation of the Ninkasi Space Program (NSP) in 2014. While NSP is an unusual project for any brewing company, it was launched with the aspiration that they could advance the science of beer production and elevate their company to a new level of innovation. According to Jamie Floyd, "Nikos and I have always loved space…. We found out through a mutual friend about a group of people who were designing the launch device for the fifth anniversary of the first amateur rocket launched into space. It was being done out of Blackrock City, where Burning Man happens in Northern Nevada. We were approached by an amateur rocket team called Team Hybriddyne who needed support for the launch. But we decided that we wanted to do more than just help sponsor the flight, but also be part of it, especially because of my love of aeronautics."[68]

Now that they had secured a launch provider, the next natural question was what type of payload they could send; that is, what payload would have benefits to their brewery? They decided that their best option was to launch yeast into space. While yeast had been sent into space in the past (such as with the experiments of Kirsten Sterrett in the 1990s), to their knowledge no space-flown yeast had ever been used to make commercially available beer.

NSP ended up partnering with two amateur launch groups, the Civilian Space eXploration Team (CSXT) and Team Hybriddyne. They worked with these partners to design a system to protect the yeast from the g-force of launch, extreme cold of space, and the heat of reentry. Through this partnership, NSP launched "Mission One" in July 2014 in the Black Rock desert. As is often the case, even with well-funded government space programs around the world, NSP ran into some technical challenges that nearly ended their space dreams. The launch of the rocket went as planned, but sometimes landing a rocket payload is just as challenging as launching it. The payload was being tracked by devices being tested by a government agency, but two out of three tracking devices failed during its descent, causing the payload to drift miles from its intended landing site. As a result, they were not able to locate their experiment for 27 days. To complicate matters even more, they were searching for the payload in the middle of a military ordinance testing area.

Meanwhile, the yeast samples had been roasting in the summer heat (approximately 108 degrees Fahrenheit) of the Nevada desert for the entire time, killing virtually all of the yeast samples. According to Floyd, "We had estimated that we had about 12 hours to retrieve the payload before the yeast would no longer be viable. I was actually surprised, [after] 28 days when we got the rocket back, there were still some live yeast cells there. They weren't viable to produce beer, but we were surprised to find even a few live cells."[69]

Floyd acknowledged that they learned some valuable lessons from this failure. While it is doubtful whether viable yeast samples would have survived that long in any circumstances, "it was a mistake to put the yeast at the nose cone tip because that area of the rocket experiences the greatest temperature change during flight, and thus makes it more difficult to keep the yeast alive."[70]

Armed with these lessons, NSP was able to launch Mission Two just a few months later, in October 2014. This time they launched with UP Aerospace of Denver, Colorado. UP Aerospace was founded in 1998, offering customer payloads relatively inexpensive access to suborbit (roughly 115 km). Some of their more publicized launches occurred when they launched the ashes of James Doohan, who played Star Trek Chief Engineer "Scotty" (incidentally, these ashes were temporarily lost when the payload drifted off course), and the ashes of Mercury astronaut Gordon Cooper.

Mission Two carried six vials of yeast and was launched from the new commercial space port in Truth or Consequences, New Mexico, aboard the SpaceLoft-9 rocket. Once again, the NSP team ran into some challenges that could have derailed their mission. Dry ice was needed as an essential component of the mission to keep the yeast sufficiently cooled during flight and recovery. They had planned on buying the dry ice at the local Walmart, but when they got there, it was sold out—and they needed to have the payload packed on the rocket at 3:00 a.m. the next morning.

In fact, there had been dry ice in stock a few hours earlier, but hunters had purchased the entire stock of dry ice before the Ninkasi team arrived at the store. Ninkasi founder Jamie Floyd scrambled to find an alternative dry ice vendor—including calling hospitals and other nearby Walmart stores—to no avail. He then learned that another Walmart had dry ice in stock, but this store was 180 miles (290 kilometers) away. Floyd would not be deterred by this and raced to Alamogordo in his black convertible Camaro rental car. Since they had to travel so close to the Mexican border, they had to pass through multiple border stops. Floyd recalled, "The border guards were used to the unusual declaration because payloads for rockets launching at the nearby White Sands Missile Range use dry ice for such purposes all the time."[71] Despite their heroic effort, they did not make it back by the 3:00 a.m.

**This sounding rocket launched a yeast experiment into space in 2014 for Ninkasi Brewing Company (Ninkasi Brewing Co.).**

deadline, but fortunately, UP Aerospace allowed them to switch launch times with another organization that morning. "They allowed NASA to fill the rocket before us—they were supposed to do it after us—which bought us another hour or hour and a half," said Floyd. "It was pretty amazing and a definitely last-minute adventure."[72]

The launch was successful and achieved an altitude of 408,035ft (77.3 miles), but since this was only a suborbital flight, the actual time in weightlessness was very limited—only four minutes. Unlike their first launch attempt, this time the payload was recovered by the NSP team in under three hours. When they returned it to their Oregon facility, they found that while the yeast samples in two of the vials were no longer viable, the samples in the other four vials had survived. Two of these vials contained an ale yeast that would be used for their new product. Two other vials had a strain used for German lagers, and the final vial contained an "alt" yeast ("alt" yeast is Altbier—or Altstadt, which means "old town" in German) strain.

Several months later they released Ground Control beer. Manufactured with the yeast that they had launched into suborbit, Ground Control was an Imperial Stout, brewed with Oregon hazelnuts, star anise, and cocoa nibs and fermented with the Ale yeast that was sent into suborbit. In a Ninkasi press statement, Ninkasi CEO Nikos Ridge said, "After almost two years of research, development, lab time, and two separate rocket launches to garner space yeast, we have finally completed our mission."[73] He added, "It was a project born out of passionate people coming together to try something new and we can't wait to share it with the world." Jamie Floyd echoed this sentiment in a later interview. "It's much deeper than the PR piece. Obviously, it's created some great storytelling, but for me, not having been able to pursue that side of it [a career in aerospace], it was really a great way for me to have that piece of it—even though it wasn't part of my actual everyday life."[74]

Passion may have been the primary motivating factor for Ninkasi, but the publicity generated may well have been the most valuable aspect of this project. The suborbital experiment and the subsequent beer have been written about in dozens of articles and was even featured (and consumed) on Neil deGrasse Tyson's *Star Talk Radio*. It was their first event that they found most humbling, however. They were able to participate in a 45th anniversary event for the Apollo Moon landings in San Diego—an event featuring dozens of astronauts and other luminaries from the Apollo program, most of whom enjoyed the Ninkasi beer brewed from space-flown yeast.

Ninkasi released a second edition of Ground Control in 2016 and a third edition in 2017. The Ninkasi team is also not discounting the prospect of a future, more ambitious project.

Ninkasi generated a substantial amount of publicity around their brand, but it was the inspirational value of their efforts that truly motivated these beer makers. They wanted not only to enable their own dreams of space exploration, but to inspire others to do the same. According to Jamie Floyd, "That's one of the best parts of the project…. If we put our minds to it, we can help be a part of that."[75] Floyd also looks with anticipation to the day that beer will actually be produced on other planets: "We're all scientists on one level or another in the alcohol business. Personally, I think it's a fascinating idea to think about the microbiological level of study when colonizing Mars…. I think there's an awful lot of amazing abilities to see how that's all going to play out."[76]

## Lunar Beer: University of California, San Diego

More than twenty years after Kirsten Sterrett's space yeast experiment, another student beer yeast experiment is now underway. In 2017, a group of engineering students at the University of California, San Diego, entered the space beer fray by proposing an experiment to determine whether beer can be brewed on the Moon. They call themselves "Team Original Gravity,"

vying for a spot to have their experiment sent to the Moon aboard the Indian
TeamIndus spacecraft.

Bioengineering student Neeki Ashari explained that "the idea started
out with a few laughs.... We all appreciate the craft of beer, and some of us
own our own home-brewing kits. When we heard that there was an opportunity to design an experiment that would go up on India's moonlander, we
thought we could combine our hobby with the competition by focusing on
the viability of yeast in outer space."[77] In the spirit of Kirsten Sterrett's space
yeast experiments, Team Original Gravity also hopes to send yeast into space,
but this time to the surface of the Moon.

According to Ashari, "Yeast is a prevalent microorganism. It is in our
food (bread), beverages, and pharmaceuticals (insulin)... Understanding
yeast viability in space may have consumptive and clinical applications for
the future of space exploration and colonization."[78]

In their experiment, the students intend to combine the fermentation
and carbonation stages of the beer production. Their fermentation vessel will
appropriately be roughly the size of a beer can. "Our canister is designed
based on actual fermenters," said Srivaths Kaylan, a nanoengineering student.
"It contains three compartments—the top will be filled with the unfermented
beer, and the second will contain the yeast. When the rover lands on the
moon with our experiment, a valve will open between the two compartments,
allowing the two to mix. When the yeast has done its job, a second valve
opens and the yeast sinks to the bottom and separates from the now-
fermented beer."[79]

As such, while lunar beer is the publicity hook for this experiment, these
students have more universal objectives than just advancing the prospect of
having a pint on the Moon. "Our experiment is much deeper than just brewing beer on the Moon," explained Ashari. "Scientifically, we are curious about
yeast and how it behaves in a lunar gravity environment. It is extremely beneficial to understand how low-gravity conditions may affect microorganisms.
Data results will leap us closer to colonization and survival on Mars."[80]

In theory, this experiment might not only help to show the possibility
of beer production on the Moon, but will investigate benefits for pharmaceuticals and yeast-containing foods.

"A majority of the biomedical experiments conducted in the ISS (International Space Station) have clinical applications for both Earthbound and
space-bound use. We study experiments in space for the benefit of mankind.
Astronauts are the eyes and ears for the scientists and engineers down on
Earth. They are explorers, looking for alternative methods and techniques to
sustain/prolong the lives of human beings. We are looking for simplified solu-

tions for an unborn generation on Earth."[81] This type of visionary approach sounds very similar to some of the key innovation disruptors of our age. Ashari seems to look to those visionaries as examples and inspiration: "To those who might say it is frivolous, it's okay to have crazy ideas. People once thought Elon Musk's idea to develop reusable rockets was frivolous. But guess what? He did it. It is because of his 'crazy' ideas, that he has now revolutionized the space, automotive, and energy industries."[82]

## Mars Desert Research Station (See also Mars agriculture experiments in chapter 6)

As will be detailed in chapter 6, if alcohol is to be produced on the Moon or Mars, the ability to grow plants is an absolute requirement; and there are organizations around the world investigating whether crops may be possible in those alien environments. One such experiment was specifically focused on the goal of manufacturing beer on Mars. A team at the Mars Desert Research Station (MDRS) conducted an experiment to determine whether beer production might be possible at future Mars settlements. MDRS is a simulated Mars base in Hanksville, Utah. This facility is operated by The Mars Society, a space advocacy membership organization devoted to the goal of human exploration of Mars. MDRS was constructed in 2001 shortly after The Mars Society completed a similar facility called the Flashline Mars Arctic Research Station on Devon Island in the Canadian Arctic. Since its construction, MDRS has hosted almost 200 crews who have performed a plethora of scientific research, mission operations exercises, and simulated Mars surface missions.

In February 2015, the 149th crew of MDRS investigated whether future Martian settlers would be able to brew indigenous beer at their new homes, as settlers have done on Earth for thousands of years. A team of researchers led by Principal Investigator Paul Bakken observed how hops and sorghum, two core ingredients in beer production, grew in Martian soil. Since no actual Martian soil has ever been retrieved from that planet, Bakken and his team used a simulated Mars soil called JSC 1A (simulant) created by the NASA Johnson Space Center based on data from the Mars Viking missions of the 1970s as well as the Pathfinder mission in the 1990s. The simulant is largely derived from volcanic ash in Hawaii.

The ability to grow crops (see chapter 6) is an absolute requirement for any permanent settlements on Mars or anywhere else. As such, Bakken sees these crop experiments as having value in understanding methods that might be employed in real Mars agriculture one day. That said, Bakken also believes that there is more to human life than just the minimum necessities. "If we plan on colonizing Mars one day, in addition to food crops and fiber crops, it's anticipated that you might want to make living on Mars a little more enjoyable and beer can certainly be a part of that as well as being an excellent way to store calories."[83]

As they planned the experiment, selection of the specific type of hops was particularly challenging. Bakken and his team wanted to select a variety of hops that was as durable as possible. They decided to use the Cascade variety of hops because it tends to be disease resistant and can thrive in a wide range of climate conditions. According to Bakken, Cascade is also the hops variety that is the most widely used in craft breweries in the United States.

As for the sorghum, they acquired a variety called "dwarf sorghum." Sorghum was chosen not only because beer can be made from it, but also because it is high in nutritional value. The dwarf variant of sorghum is also advantageous since fully-grown plants are only half the size of other sorghum varieties—roughly [2–3] feet tall. The smaller stature of this crop will probably be better suited within the limited space that will likely exist in future space habitats.

Time was not on the side of Crew 149, however. Their mission was limited to two weeks in duration. Because of this, they arranged for subsequent crews to continue tending the plants so that the overall experiment had a

Initial plantings in simulated Mars soil at the Mars Desert Research Station (MDRS) in Utah in 2015 (Paul Bakken).

five-week duration. This was still not enough time to observe a full growth cycle, but enough time to collect valuable data. The team seeded both the hops and sorghum in pots filled with the Mars soil simulant and also planted a control group in regular potting soil. Despite the time limitation, they were confident of success. According to crew member Kellie Gerardi, "We were cautiously optimistic that we would see germination—and we did. I wouldn't say we were surprised, but we were certainly excited."[84] Both crops grew well and began the initial phases of their secondary stages of growth. "The interesting finding was that both species actually did a little bit better in the regolith simulant than in regular soil."[85]

While these findings were certainly promising, they did not show how well the plants would perform during a full growth cycle within the Mars simulant. As such, Bakken conducted some follow-up experiments after leaving MDRS. In these studies, the crops did not do as well later in their growth phases. Bakken attributes this largely to lack of nitrogen in the Martian regolith—and neither of the plant species are ones that can extract nitrogen from the air. As a result, after the initial growth stage, their growth was stunted and the plants became sickly.

When Bakken supplemented the soil with nitrogen, the sorghum did much better, but the hops did not do as well. Bakken concedes that this might have been a result of other factors, specifically the cramped indoor space the crops were being grown in.

Bakken would like to see follow-up experiments conducted to measure the nutrition content in these crops. "It's possible that you could have a plant that looks perfectly fine, but really wouldn't be that good for you."[86] One of the factors driving this concern is that some toxic chemicals have been detected on the Martian surface. Perhaps the most concerning of these chemicals is the existence of perchlorates. Perchlorates are naturally occurring chemicals that consist of "one chlorine atom bonded to four oxygen atoms."[87] It is often used as an oxidizer in rocket fuel and can be toxic to plants and animals in large enough quantities. It can be absorbed into plants.

If the crops prove to be unhealthy for human consumption, Bakken believes there are other ways to eliminate the perchlorates prior to growing crops in the soils. This includes super-heating the soil to remove the perchlorates.

Even with all the challenges that will face initial Mars explorers, Bakken remains confident that not only will they produce beer, but it will happen early in the settlement process. "I can't think it would take too long for people to produce beer or some other alcohol wherever people go. Even if it's not an 'official' crop intended for beer production, they tend to want to bring their creature comforts with them—and alcohol is certainly one of them."[88]

The knowledge gained from growing hops and sorghum have benefits well beyond beer. According to Kellie Gerardi, "As exciting as the concept of beer on Mars was, the project was fundamentally about the ability to grow crops in a Martian soil simulant."[89] Without the ability to grow crops, humans will not be able to establish long-term settlements in space.

# 5

# Wine, Whiskey and Innovation

## ARDBEG: The First Distillery in Space

One of the most unusual entries in the space booze business was Scottish single malt whiskey producer Ardbeg (Scottish Gaelic—*An Àird Bheag*, meaning "the small promontory").

Located on the island of Islay (pronounced: eye-la), Ardbeg has been in existence for over 200 years, beginning commercial distillation in 1815. For most of Ardbeg's existence, however, it produced whiskeys that were used in blends by other producers rather than for the strong single malt product that it is known for producing today. It continued producing for the blended whiskey market well into the twentieth century, but its story almost ended in 1981 when the Ardbeg distillery closed. Ardbeg got a new lease on life in 1987 when it was purchased by Hiram Walker.

The true rebirth of Ardbeg occurred, however, when it was purchased by Glenmorangie Company (now part of Moët Hennessy) in 1997 and began full production. This time Ardbeg produced single malt whiskey that became known as one of the peatiest (or smokiest) single malt whiskeys in the world. The Ardbeg 10 Year Old became a staple of that company's growing reputation.

However, it was random chance that thrust Ardbeg into the forefront of space drinks research.

According to Jeffrey Manber, the co-founder and CEO of Nanoracks, a company that launches goods and services to low Earth orbit, "A colleague of mine, who's a co-founder of Nanoracks, was at a whiskey tasting in Rye, New York. Apparently, he met some folks associated with Ardbeg and when they learned what Chris does for a living, they got very intrigued, which began a series of discussions."[1] In their early discussions, Ardbeg expressed an interest in sending a bottle of Ardbeg to the International Space Station, but Manber explained that this would violate NASA rules. Any experiment that they proposed would need to be part of a public awareness effort or have real scientific value. After about a year of conversations, they finally came up with a viable project involving terpenes. Terpenes are volatile unsaturated hydrocarbons and are "the building blocks of flavour for whisky spirits as well as for many other foods and wines."[2] They can be found in many varied products, including cannabis and paint thinner.

Although these discussions had been proceeding for several months, learning of the prospect of a space experiment still came as quite a shock to some people at Ardbeg. According to Global Brand Ambassador David Blackmore, "Myself and colleague Hamish Torrie both received voicemails and emails over a weekend, and initially both thought that it was some sort of prank, as we couldn't believe that anyone would be calling us to discuss launching Ardbeg onto the ISS. However, our Director of Distilling and Whisky Creation, Dr. Bill Lumsden, had spoken to the Nanoracks team, and realized that the project was very serious and that we had quite an Ardbeg fanclub at Nanoracks. From there, it was a matter of a day or so until we agreed to the collaboration."[3]

NASA had never done an experiment to determine the impact of microgravity on growth of terpenes, and since terpenes had utility for other products as well, including perfumes, cosmetics, and food additives, it seemed to be a compelling first experiment for Ardbeg. Nanoracks and Lumsden were able to gain approval from NASA to conduct the experiment.

Thus, Ardbeg thrust the whiskey world into the space age when their terpene experiment was launched to the International Space Station on a Russian Soyuz vehicle from Kazakhstan in September 2011. While discussions had gone on for a year, their window for designing the experiment prior to launch was extremely short, so they scrambled to assemble an experiment that was viable but could also produce meaningful scientific knowledge, and possibly provide insights that could impact the future of whiskey making.

One significant challenge was how to create an authentic aging experiment. Quite obviously, they would not be able to send oak barrels to the ISS, but the Lumsden and Nanoracks teams came up with a novel approach to address this. A small quantity of Ardbeg distillate, as well as "oak wood shavings from the inside of a charred American White Oak ex–Bourbon barrel,"[4] were sent to Nanarocks in Houston. Once there, scientists packaged them in MixStix™ (containers that are used to mix samples in microgravity). As such, some MixStix™ samples were sent to the ISS, and a control sample remained in Houston to be compared to the "flown" sample when they were returned from space.

The experiment lasted for 971 days aboard the ISS before the distillate was returned to Earth on September 12, 2014. When the samples were examined in the lab, it became clear that aging in the space environment did have a noticeable impact on the taste of the final product. According to Bill Lumsden, "When I nosed and tasted the space samples, it became clear that much more of Ardbeg's smoky, phenolic character shone through—to reveal a different set of smoky flavours which I have not encountered here on earth before."[5]

Jeffrey Manber from Nanoracks also instantly noticed a difference in the aroma of the space-flown samples, stating, "The ones in the microgravity environment had a deeper, richer color…. It was a really deep, rich

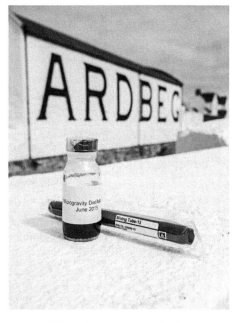

Ardbeg distillate that flew to the International Space Station in 2011 (Ardbeg).

smell. It stood out from all the other containers that returned from ISS since ours had this wonderful smell."[6]

After a thorough examination of the space-flown whiskey samples and the control samples that remained on Earth, Lumsden released a white paper in 2015 detailing the results of the experiment. According to his white paper, the ISS samples contained "a very focused flavor profile, with smoked fruits (prunes, raisins, sugared plums and cherries), earthy peat smoke, peppermint, aniseed, cinnamon and smoked bacon or hickory-smoked ham. The aftertaste is pungent, intense and long, with hints of wood, antiseptic lozenges and rubbery smoke." As for the aroma, it was described as "intense and rounded, with notes of antiseptic smoke, rubber, smoked fish and a curious, perfumed note…."[7]

David Blackmore recalled the difference more colorfully during a panel discussion held by Explore Mars, Inc., in Washington, D.C., in 2017, stating, "There was a difference—quite a significant difference. The sample that had been on ISS was quite different—much more intense in character. Ardbeg is very smoky—so the regular Ardbeg sample was just as you'd expect. The sample that had been on ISS was much more intense."[8] According to Ardbeg, it had notes of smoked fish, seaweed, and smoked ham—with several congeners that had never been found in whiskey before.

Blackmore suggested, however, that additional testing will need to be conducted to verify whether the difference between the samples that flew in space and those that remained on Earth were in fact a result of microgravity or a result of other factors that were unique to the "flown" sample. Blackmore was concerned that it might be premature to suggest "that all the effects were caused by microgravity when one sample sat in the tranquility of the Ardbeg facility and the sample that went to the ISS had to go through reentry and was essentially shaken like a cocktail for an hour solid."[9] If Ardbeg decides to conduct a future test, they will want to design an experiment that is more robust and can compensate for the extreme level of shaking that occurs during launch and landing—or find ways to subject the ground samples to any approximation of the shaking that the space samples will be subjected to. Blackmore concluded by admitting that while they "did learn that the spirit that spent time on the ISS differed measurably and through human olfactory analysis," they were still not entirely sure why this happened. "The next step of course is to investigate why."[10]

Regardless, Ardbeg was able to take full advantage of the notoriety they gained by becoming the first scotch producers in space. According to Jeffrey Manber, "Don't let their stodginess—their 200-year-history fool you, they were brilliant at taking the growth of terpenes and leveraging that into an entire, sophisticated marketing campaign both in Europe and the states. They were remarkably successful, and we've begun discussions about flying again and testing more."[11]

The Ardbeg publicity campaign turned out to be just as non-traditional as the space flight itself. Ardbeg's advertisements were whimsical and offbeat and would have looked at home in Silicon Valley or even in a Monty Python sketch. Ardbeg also did not wait for the experiment to return to Earth to take advantage of the notoriety of being the first scotch producer in space. In May 2012, Ardbeg began The Ardbeg Rocket Tour that featured a smoke billowing "life-size" (30 foot tall) rocket that traveled to cities around the United States. In an Ardbeg press release, Ardbeg U.S. Business Director Geraud Leclercq stated, "This tour is the perfect opportunity to educate our consumers about this exciting Ardbeg maturation experiment in microgravity, which holds out the tantalizing prospect of, maybe, being able to develop innovative new products in years to come."[12]

Continuing with the space theme, in 2012 Ardbeg also introduced a new 12-year-old single malt whiskey called "Galileo." In one of the advertisements for this scotch, a rocket resembling an Ardbeg bottle launches from the moors of Islay, Scotland, and lands at a Moon base. A narrator then concludes by saying, "Ardbeg has conquered Earth. Next step, the Moon?" Most recently, in 2015 Ardbeg released the final edition of their premiere product, Supernova, to commemorate their space mission and also celebrate the 200th anniversary of the founding of the company. Ardbeg had established itself as arguably the "peatiest" on the market; Supernova was intended to elevate that distinction to new levels. In the description on the back of the bottle, it states, "In theory, SUPERNOVA is one of the peatiest whiskies on Earth. Yet, to taste, it is also intensely spicy and unfathomably sweet. Truly a mystery of the universe...."

Having achieved success with the initial microgravity experiment, Ardbeg hopes to conduct additional, even more ambitious space missions in the upcoming years. Ardbeg is likely to take advantage once again of the expertise of Nanoracks to enable the next experiment in space, but according to Jeffery Manber in a 2018 interview, "I'm not privy yet exactly—but something robust and using an oak container. I think they want something a little more realistic for what happens in their industry. I mean, we can't send up a barrel, but something more we can do to test the understanding of what happens."[13]

Regardless of what the scotch experiment turns out to be, Manber is excited about the prospect of pairing with Ardbeg again. "It's hard to find companies willing to be pioneers. To have a partner like Ardbeg that is willing to make this sort of commitment augurs well for the future of commercial space research into flavourings and what it changes for consumer products in general."[14] In fact, whiskey producers may be uniquely suited to partner on space projects. Manber recalled someone at Ardbeg explaining that they are a 200-year-old company that makes a product that takes at least 10–15 years to produce. They are currently working on products that will not be released until the early 2030s—the timeframe that NASA hopes to go to Mars. "So, we're actually very comfortable with the space community because it moves at a pace much like we would."[15]

## Suntory Single Malt Space Whiskey

Ardbeg's monopoly on space whiskey did not last long. In 2015, Japanese whiskey maker Suntory launched their own whiskey experiment to the International Space Station (ISS). Like Ardbeg, Suntory has a long history. Suntory Holdings Limited is one of the oldest alcohol distribution and manufacturing companies in Japan. Founded in 1899 by Shinjirō Torii, it began as a store that sold imported wines, but in 1923 started the first whiskey distillery in Japan, the Yamazaki Distillery. The new distillery produced a single malt in the Scottish whiskey-making tradition. The first bottles of Suntory Shirofuda

(White Label) were sold in 1929 and were the first true Japanese whiskey. This being the case, the Suntory website states, "The Yamazaki Distillery is thus the birthplace of Japanese whiskey. Nestled proudly on the periphery of Kyoto, this region was formerly referred to as 'Minaseno,' where one of the purest waters of Japan originates."[16]

Only interrupted by World War II, Suntory expanded throughout the twentieth century, introducing numerous new whiskey labels and expanding into many other drinks and products. They also extended their worldwide reach, becoming popular in North America and Europe and even finding their way into popular culture. Perhaps most memorable was in a scene of the film *Lost in Translation*. A tuxedoed Bill Murray is facing a translation problem while filming a Suntory commercial as he continually turns to the camera with a glass of whiskey in hand and says, "For relaxing times, make it Suntory time."

Suntory has always been a company that has valued innovation, and in this spirit, they conceived of an experiment to deliver samples of their whiskey to the ISS to observe "development of mellowness in alcoholic beverage through the use of a microgravity environment."[17] The Suntory experiment launched in August 2015 on a H-II Transfer Vehicle No. 5 (also known as "Kounotori5" or HTV5) from JAXA's Tanegashima Space Center headed to Japan's "Kibo" module on board the ISS.

In fact, the initial idea was conceived in 2011, around the time that Ardbeg launched their experiment to the ISS, but it took Suntory several years to develop their own concept. According to a Suntory press release in June 2015, "Our company has hypothesized that 'the formation of high-dimensional molecular structure consisting of water, ethanol, and other ingredients in alcoholic beverages contributes to the development of mellowness.'"[18] The experiment consisted of two groups of whiskeys that were contained in glass flasks. The whiskeys ranged in age from recently distilled whiskey upwards to twenty-one years old (new, 10 years, 18 years, and 21 years). One of these groups resided on the ISS for a period of a little more than a year while the other group of samples still remains on the ISS (as of early 2019).

None of the samples would be consumed in space, but when Canadian astronaut Chris Hatfield read about the experiment, he comically tweeted, "Whisky in space—this will truly demonstrate the discipline & self-control of astronauts."[19]

Suntory's mission has valid scientific goals that could impact future products as well as the general understanding of the aging process of whiskey and other alcoholic beverages. Nonetheless, they were clearly aware of the tremendous amount of publicity that the mission was generating and how it could impact their corporate image around the world. As whiskey expert Tom Fischer noted, "Scientifically, I'm sure something will happen … but more than anything else, it will be the romance, the interesting story behind drinking bourbon that was in space—the subjective mind will automatically think it's better."[20]

According to Suntory, they have no current plans for future space experiments. However, since they still have samples aging on the ISS, it is hard to say what their future plans will be depending on the results of their experiments.

## Space Bubbly: Bringing Conviviality to Space

Consuming carbonated beverages in space may be an obstacle, but a famed producer of what is considered the most "elite" form of carbonated alcoholic beverages believes it is an obstacle well worth conquering. Champagne producer Maison Mumm has partnered with a French design company called Spade to produce a bottle that will enable space tourists and astronauts to consume champagne in space and assure that it is a satisfying experience—without the typical side-effects of consuming carbonated beverages in space. According to the Maison Mumm website, "Playing with the constraints of zero gravity. New sensations. New aromas. A new way of raising a toast and bringing a human touch in space. Designed under the close supervision of Maison Mumm's Cellar Master, Mumm Grand Cordon Stellar is filled in the House's cellars in Reims and meets all its rigorous expectations in terms of quality."[21]

Maison Mumm has been building that tradition of rigorous expectations since their founding in 1827. Since then, they have taken great pride in their ability to innovate to assure a high level of quality, inspiring the motto, "Only the best." As such, their decision to pursue space champagne was not as unlikely as it outwardly appeared.

However, the origin of this concept did not originate with Mumm. It came several years earlier, when Spade head designer Octave de Gaulle was attending l'ENSCI, a French design school. He undertook a school project that would lay the groundwork for Maison Mumm's space champagne and bottle. "It actually started when I made a bottle of wine for zero G for my graduation project as a designer in 2013,"[22] recalled de Gaulle. He wanted to create a product that would "not just bring the substance (of alcohol), but also bring the conviviality and the experience of sharing and tasting, which is not easy to achieve when you're not under the laws of gravity anymore."[23]

De Gaulle had been inspired by a 1950s comic called *The Adventures of Tintin: Explorers on the Moon,* in which one of the characters is challenged by the prospect of drinking a glass of whiskey in microgravity.

De Gaulle believes that alcohol consumption will be very important in space—for very positive reasons. Champagne and other alcoholic beverages can play an important role in the psychological wellbeing of people who operate in isolation. As tourism and settlements in space begin, social consumption of alcohol can play an important function with the interactions between these pioneers. "As a Frenchman, we use wine and champagne to enhance social interactions."[24]

In 2015, de Gaulle partnered with Mumm and after three years of design, Mumm Grand Cordon Stellar announced their space champagne in September 2018. They describe this space bubbly as "a groundbreaking feat of technology that makes it possible for astronauts and other space travelers to enjoy Champagne in the challenging surroundings of zero gravity."[25]

Mumm knew that one of the biggest challenges is getting the liquid out of the actual bottle. To solve this problem, the Mumm team harnesses the gas within the champagne to "expel the liquid into a ring-shaped frame, where it is concentrated into a droplet of bubbles"[26] that "can then be passed to someone and released into the air, where it floats until gathered up in a specially designed glass."[27] Those droplets then converge into "an effervescent ball of foam."[28]

In this form, it creates an unusual effect for the spacefaring drinker. The "effervescent ball" of bubbly returns to a near liquid form when it enters the consumer's mouth. Mumm's Master Distiller Didier Mariotti describes, "It's a very surprising feeling … because of zero gravity, the liquid instantly coats the entire inside of the mouth, magnifying the taste sensations. There's less fizziness and more roundness and generosity, enabling the wine to express itself fully."[29]

As they investigated which champagne would be most suitable for microgravity conditions, de Gaulle recalled that, "with the champagne itself, the first we tried was a version that was less carbonated,"[30] as they had heard about the problem of moist belching and stomach discomfort in space. However, they reconsidered and opted for a champagne with a standard level of carbonation. As de Gaulle explains, "Considering the time that you actually spend in micro-g and the content of this bottle—it is only half a bottle—you only drink the tiniest bit and it is incredibly intense in terms of flavor … you have this capacity to layer everything in your mouth, so you drink very little. So we went back to the regular champagne and we didn't change anything in the recipe."[31]

In keeping with French sensibilities, style and functionality are not mutually exclusive. As such, to add a more authentic feel to champagne toasts in space, the space bubbly team also designed a special glass that has a "tapering stem, no base and a small cup to catch the floating droplets"[32] that allows microgravity consumers to drink and toast in a more traditional fashion.

To test their bottle of champagne and the glass, they utilized the services of a European zero-gravity flight service called Air Zero G that is run by Novaspace (a subsidiary of the French space program CNES) and French airline broker, Avico. Like the Zero Gravity Corporation in the United States, Air Zero G creates weightless conditions by flying parabolas, but in their Airbus A310.

Mumm did not want this flight to be considered to be a stunt. To assure a successful test and to add credibility to the process, the Maison Mumm team recruited experienced crew members. According to de Gaulle, "We did this experiment with very experienced people like astronauts and the crew of the plane who know the effect of zero gravity … but somehow, the day I brought the champagne and all the professional staff who are used to the magic of zero gravity—they were really fascinated by the experience. For me as a young designer, it was quite cool to give these people a new experience/perception in zero-G."[33]

One of the crew members on their flight was former ESA astronaut Jean-Francois Clervoy, who appeared to be thoroughly enjoying this particular microgravity experiment. He stated while drinking from the glass during a weightless part of the flight, there is "no need for a table. The guest simply floats—captures globes of champagne. It's magical!"[34]

In the short term, Maison Mumm hopes that their product will be served regularly on zero-gravity flights, but their ultimate goal is for it to be consumed on commercial space flights when they begin. Clervoy emphasized this vision when he said, "In coming years, private companies will fly into space and will enjoy champagne in zero-G or on the Moon—maybe one day on Mars."[35]

Their second zero-gravity flight took place in September 2018, flying out of the French champagne region. This would be the "official" rollout of Mumm Grand Cordon Stellar, so the press was also invited aboard. To add additional excitement to the event, Mumm also invited Olympic sprinter Usain Bolt to participate in the flight, during which he attempted a sprint in reduced and microgravity conditions as the other passengers toasted him with the Maison Mumm champagne.

With this success, they also hope to inspire more elements of design and style to be integrated into future space plans. As Octave de Gaulle explains, "For the past 40 years, space travel has been shaped by engineers rather than designers. Instead of seeing zero gravity as a problem to be solved, we look at it as a design possibility."[36]

Commercial flights into space may not be available quite yet (2019), but interested customers can buy Mumm Grand Cordon Stellar now. It comes in a metallic, space-age carrying case that contains the specially designed bottle of champagne as well as two of the tempered-stem glasses. However, at a price of 25,000 euros, it costs the same as a moderately priced car, but people who buy a bottle are eligible for a chance to sip Mumm Grand Cordon Stellar on a zero-gravity flight.

Since their September 2018 rollout, the Mumm team has conducted some audacious publicity stunts showcasing their space champagne ambitions. In October 2018, they unveiled a Space Odyssey-themed marquee at the Birdcage at Melbourne Cup in Australia. This featured a twelve-meter-high rocket called the Mumm Rocket as well as the Mumm Space Odyssey Orchestra that was composed of professional Australian orchestral musicians. They also teamed up with Martin Hudak, the former Senior Bartender from the Savoy in London, to create a cocktail called "The Mumm Walk" cocktail, named after the apocryphal "Moon Walk cocktail" that was allegedly served to the Apollo XI crew when they returned from the Moon.

## Wine in Space

> "Scientists need inspiration, and with inspiration, you need wine. Once we take wine to Mars, everyone will follow."
> —Ramaz Bluashvili, TV space presenter
> in country of Georgia[37]

When French astronaut Patrick Baudry brought a bottle of vintage 1975 Château Lynch Bages into space on the Space Shuttle in 1985, it was to honor his French culture. (Note: The bottle remained unopened.) This was only intended as a symbolic gesture at the time, but it also demonstrated that when humans start settling new worlds, they will likely demand beverages such as wine. Space experiments specifically focused on wine production have not been as numerous as for other alcoholic beverages, however. This is likely a result of the fact that wine production presents more challenges than most other alcoholic beverages. Even the Mumm champagne project is not trying to produce champagne in space but select terrestrially produced bubbly that is suitable for consumption in that environment. Nevertheless, there are some space-based wine experiments underway, not in France or Napa Valley as one might expect, but by China's massive and growing wine industry.

In September 2015, China launched Cabernet, Merlot, and Pinot Noir vines to their Tiangong-2 or "space palace" space station. The experiment was designed to test grape vine resistance to drought and cold weather; an experiment with many practical applications on Earth. This is extremely important to the Chinese wine industry; China has a quickly growing wine-consuming population. According to *The Guardian*, "The Asian giant now consumes more red wine than any other country and has more vineyards than France. Estates are popping up from the frosty northeastern province of Liaoning to the scorching deserts of Xinjiang."[38]

To help enable this surging demand for wine, the Chinese wine industry hopes to find innovative ways to better utilize some of their harsher environments that tend to be colder and are more arid than is typically the case in good grape-growing regions. This includes the Gobi Desert and the Ningxia region of China. Although Ningxia is considered one of China's highest quality wine regions (often called China's Bordeaux), wine growers regularly need to bury their vines at night to protect them from cold weather. With this in mind, the vines selected for the mission to the "space palace" came from the Helan Mountain east region in Ningxia from a nursery owned by the Chenggong Group.

The vines were scheduled to undergo thirty days of experiments in space. Researchers hoped that growing vines in space will cause mutations that will enable wine grape vines to grow and thrive in extreme conditions. It is unknown what the results have been from these tests or what future plans China has for using space to help their wine industry, but it is something that wine producers in the rest of the world should watch closely.

## Wine on Mars

*"I look forward to enjoying my space-born wine. Wine born in space."*

—Bill Nye[39]

In 2019, a project was announced in the country of Georgia (a former republic of the Soviet Union) with the goal of one day producing wine on Mars. The IX Millennium Project will be looking at the impact of radiation, dust, lower atmospheric pressures, and other challenges on the growth of grape vines. In addition to the potential for wine in future Mars colonies, the researchers of this project hope to make an important contribution to Mars agriculture.

However, as a country that claims to be the birthplace of wine—viticulture dates back over 8,000 years in that country—project leaders thought it was a perfect way to contribute to the future of space exploration. According to Nokoloz Doborjginidze, who is founder of Georgia's Space Research Agency and a researcher on the Mars wine project, "Our ancestors brought wine to Earth, so we can do the same for Mars."[40]

The project started in 2019 and plans to create a vertical garden inside a hotel in Tbilisi to investigate which types of vines will thrive through this type of agricultural environment. They also are working with Tbilisi Business and Technology University (BTU) to conduct a series of growth tests in simulated soils. Part of this testing regime will include producing a more Mars-like environment, including temperatures well below freezing, an air pressure that is equivalent to 20,000 feet on Earth, and exposure to radiation.

An artist's depiction of a bottle of wine and glasses at a restaurant on a space station of the future (Bryan Versteeg/SpaceHabs.com).

One of the challenges will be to select a robust variety of grapes that will be able to survive in these types of conditions and will be suitable for wine production. One of the varieties is called rkatsiteli, which is described as "a robust and common wine high in acidity with hints of pineapple and fennel and a fiery kick."[41] Even though they are selecting grapes for potential wine production on the red planet, the IX Millennium team thinks that white wines might be more suitable. "Whites tend to have more resistance to viruses," explained Levan Ujmajuridze, who is the Director of the Scientific-Research Center of Agriculture in Tbilisi. He speculates that these grapes will do well in radiation because "their skin could reflect it."[42]

Whether the IX Millennium team is successful at introducing winemaking on Mars or not, their research could unquestionably contribute to numerous projects around the world to enable a sustainable future for humanity in space.

## *Private Vines in Space: The Flight of Greg Olsen*

While large American wine companies have not engaged in the wine space race, experiments have been conducted in space by at least one American—but with mixed results. In 2005, American entrepreneur Greg Olsen became the third private astronaut to pay to visit the International Space Station. At the time, he was Chairman of Sensors Unlimited, Inc., a company that developed optoelectronic devices. One of their biggest customers was NASA. In addition to Sensors Unlimited, Olsen also owns a South African winery called Olsen wines that is located in the Klein Drakenstein Mountains in Pearl, South Africa, which is an area known for premium wines.

Olsen's passion for wine production could not be contained on Earth. During Olsen's voyage to the ISS, he carried two Pinotage root stocks (small wine plants that are grown into vines). "The idea was that I could make wine that had its origin in space,"[43] commented Olsen. If he succeeded in growing wine stocks in space, over time he would be able to produce wine from grapes grown from those plants, becoming the first person on Earth to produce wine from space-flown vines. Unfortunately, the root stocks did not survive the entire trip. Olsen knew that they had survived on the flight. He had successfully kept them alive using the florescent lights on the ISS, but somewhere in the handling after leaving the ISS, they died.

For obvious reasons, Olsen thought that this would be a tremendous marketing opportunity. His wine company had previously released some terrestrial grape vintages with space themes, one called Soyuz and the other one called Orbit, but wines that were produced from grape vines that had been in space would be unique to the marketplace. From a wine marketing perspective, the trip was not a compete loss, however. In addition to the root stocks, he also brought up 500 Olsen wine labels on which were written "This label was carried into space by Greg Olsen."[44] Olsen explained, "For very special people, I put the label onto a bottle and said that while the bottle didn't go into space, the label did. And they all had the stamp of the ISS on them."[45]

Even though Olsen failed with his wine stock experiment, he is confident that wine and other alcoholic beverages have a future in human space activity. He stated, "Alcohol is a part of our daily life. When space gets to be more than just a scientific experiment, we'll want to improve food and improve living—and alcohol is part of that."[46]

As one of the few true space tourists—spending over $20 million for his

trip to the ISS—he is well qualified to comment on whether space tourism could stimulate open and accepted drinking in space. According to Olsen, "If these people are not charged with flying the vehicle,"[47] it would certainly be appropriate for space tourists or others to have an occasional drink.

**Private astronaut Greg Olsen displays a vial containing wine root stocks that he brought to the International Space Station in 2005 (Greg Olsen).**

Greg Olsen is quite optimistic that wine will one day be produced in space but thinks that beer is most likely to be the first alcoholic beverage produced because it is the least complex to make.

Regardless of whether or not his root stocks survived their space voyage, Olsen does possess a winemaking distinction related to space. He is without question one of the few (perhaps only) commercial wine makers who have been in space himself.

## Cheers: Building a Cocktail Glass for Space

> "Whatever the current cutting-edge mode of transportation was—whether it was trains or oceanliners or airplanes or briefly dirigibles—there have always been cocktails on board and always will be."[48]
> —Jeffery Kluger, Time Magazine

Virtually everyone has seen videos of astronauts in weightless conditions squirting water directly into their mouths or into the air, causing it to form large beads of fluid that float and ripple in the near-weightless environment. The astronauts open their months and literally grab the liquid from the air with their mouths and swallow. While this is a workable system to assure proper hydration, it is not necessarily the best method to fully enjoy an alcoholic beverage such as wine, scotch, or beer. Noted science communicator Bill Nye is quite confident that microgravity drinkers will not be satisfied with just a squirt bottle. "Hardcore oenologists obsess over the style of glass … so I can easily imagine there will be a whole art, a dark art associated with the proper drinking vessel in space."[49]

Companies like Vostok and Mumm have begun addressing this by producing special bottles and glasses from which to drink their bubbly. However, they are not alone in the mission to find better and more satisfying ways to serve adult beverages in space.

Cosmic Lifestyle Corporation designed and built a glass that would be able to contain a liquid in microgravity, but would also allow a consumer to sip their cocktail similarly to how they would on Earth. Such a product could play a significant role in elevating the experience of imbibing a cocktail (or other drinks) in space. Private astronaut Richard Garriott was an advisor on this project and stated, "When the folks at the cocktail glass [project] first contacted me, I immediately saw the logic of it. When you think of drinking alcoholic beverages on Earth, people put them in different glasses for different reasons.… It truly does add to the experience of [drinking]. Humans derive pleasure from the eating experience. First with your eyes, then with your nose, and then only the last step with the taste buds in your mouth. Presentationally, finding ways to make consumption more pleasant is a worthy aspect as we begin to live more long-term in space."[50]

The project was largely stimulated by the growing space tourism industry. Cosmic Lifestyle Corporation founder Samuel Coniglio noted that "space tourism is just like any other kind of tourism. It is adventure travel, and when you've gotten used to the microgravity environment, you are going to want to celebrate and party. Cosmonauts have been drinking alcohol in space since the beginning of space flight. NASA and ESA astronauts have to be very discreet and turn off the cameras."[51]

Initial planning for the glass started in 2005, but the concept was not patented until 2014 when the company was incorporated. Coniglio recalled, "We learned a lot about materials that are hydrophobic that could weaken the surface tension. Also, fruit juices have a higher viscosity than alcohol, and mixing them together in various combinations messes with surface tension as well. Finally, V-shaped grooves at 45 degrees or less have a better chance holding the liquid."[52]

Using this core design element, they came up with a microgravity cocktail glass as well as concepts for other cups and glasses. It was designed to

**Members of the Cosmic Lifestyle Corp. team in front of the zero-gravity cocktail glass. From left to right: Brent Heyning (fabrication), Russell Davis (bartender celebrity/CEO), Paul Fuller (CFO), Samuel Coniglio (inventor/COO), Nick Donaldson (designer) (Samuel Coniglio).**

resemble the profile of a traditional cocktail glass, but had special angled grooves built in that would help the flow of liquid in a microgravity environment. Since pouring liquid into a standard glass is not viable in near weightless conditions, the liquid is injected at the bottom of the glass.

Once they had successfully designed the glass, the next step was to find financial backers for the project. According to Coniglio, "We contacted about a dozen liquor companies. We were very close to building a relationship with Grey Goose Vodka, which became the 'Official drink of Virgin Galactic.' Unfortunately, the timing was very unfortunate: SpaceShip Two broke up in mid-air, killing the pilot. That shut down everything with Virgin Galactic."[53] They also had some interest from a scotch whiskey company, but that deal fell through as well.

Finally, they launched a Kickstarter campaign in 2015. The crowdfunding website stated, "The Zero Gravity Cocktail Project is a fluid dynamics and lifestyle experience design experiment. We are creating an open air drinking container that allows you to enjoy the aroma of the drink, yet keep the fluids under control. Your mouth completes the connection like a straw and you can suck the drink into your mouth."[54]

While the Kickstarter campaign fell far short of its intended funding goals, the campaign generated a considerable amount of press around the world, including appearing in the monologue of *Late Night with Seth Meyers* when he comically asked, "Our astronauts are drinking? Guys, the first step is admitting to Houston that you have a problem."

## *Weightless Scotch Glasses*

Cosmic Lifestyle is not the only company designing microgravity glassware for spirits. A group called the Open Space Agency spearheaded the Ballentine's Space Glass Project. This project set the goal of designing and building a whiskey glass suitable for use in microgravity, integrating design into space exploration. According to Open Space Agency founder James Parr, "We wanted to determine whether or not there are real aspects of design that would be making a contribution"[55] to space exploration and settlement.

Even though style and design were of paramount importance, Parr and his team initially approached the problem as a rocket engineer would, the problem being to determine how fluids will flow in space, except in this case fluid dynamics are not being used to assure that a space vehicle remains operational, but rather to create a functional glass that also satisfies the olfactory and aesthetic needs of a consumer. In a microgravity environment, this comes with many challenges. "Gravity is such an overwhelming force in our lives, we don't notice some of the other forces that can be exploited such as 'Van der waals interactions,'" commented Parr. "Thinking through the physics that's required to handle liquids. Things that are counterintuitive. For example, you can't [suck] the way you might here on Earth, because [you] ... create a vacuum and compress the air around the liquid."[56]

The glass uses a spiral convex base plate that is made of stainless steel, coated in rose gold. This plate holds the whiskey down in a reservoir at the bottom of the glass. A helix is connected to the reservoir and, as described in an article on Medium.com, "a small channel runs up the side of the glass to carry the liquid up to the rose gold mouthpiece, where the liquid waits for the space traveler to drink it. The main body of the glass is made of 3D printed thermoplastics (PLA) and is a size comparable to terrestrial whiskey glasses."[57]

According to Parr, this was an important requirement for their glass. "I think the things we were looking at were primarily about having that glass in your hand. Letting the heat of your palm warm the whisky, and the action of rolling it around the base of the glass to release the odours and smell; that olfactory experience which is a big part of the whisky drinking experience was very important to us."[58]

One of the benefits of this design is that modifications can be made to the glass to expand its functionality. For example, if a heating element is added to the base and the temperature is controlled digitally, it opens up new prospects in connoisseurship. Parr is intrigued by the possibilities, stating that "as soon as you start talking about connoisseurship, plus digital technology, it gets really exciting because you can start thinking about the mixing ratios. Since the space glass loads from the base, you can really imagine a scenario where your specific taste could be programmed into that. Imagine going up to the bar and your glass is sitting on the bar—it actually has a magnet on it—and your drink is made completely to order."[59]

Once the glass was ready, the team had hoped to test their concept in parabolic flight with ZERO-G, but the ZERO-G plane had some mechanical problems. Instead, they tested their final design at the ZARM Drop Tower in Bremen, Germany. This 146-meter-high tower (120 meters of actual drop) is located at the University of Bremen. To assure there is minimal drag within the tower, high performance pumps are used to create a vacuum prior to each drop test.

For the whiskey glass, they wanted to see which base design was more effective at holding the whiskey down at the bottom the glass, a concave design or a convex design. While they hypothesized that the convex design would prove the winner, they would not know for certain until they tested

**Ballantine space scotch glass being filled. The glass was designed by the Open Space Agency (Open Space Agency).**

it. The drop tests lasted a little over seven seconds in duration and they learned that their hypothesis was correct. The convex base was the most successful and the whiskey also flowed up the helix to the mouthpiece. While this was a very short duration test, Parr was quite satisfied with the results. He and Ballentine believe they have a glass that could be used in the microgravity of low Earth orbit.

However, the glass was not the only innovation to come from this partnership. According to Parr, "We worked with the Ballentine guys to actually design a whiskey that's tailored for drinking in space. They made a very, very short run of this whiskey which we helped to design."[60] The primary challenge they wanted to overcome was, as indicated elsewhere in this book, the fact that many occupants of space report a dulling of their taste buds—almost as though they have a cold. As such, astronauts compensate for this by adding hot spices or sauces to their food to give it an additional taste. "We applied this philosophy to the blend of the whiskey,"[61] explained Parr. They made sure that this particular whiskey had a lot of punch to overcome any reduction in the ability to taste. Ballentine's Master Blender Sandy Hyslop described it well: "In space we know astronauts miss the taste of home, so adding a fruitier, more concentrated and more floral blend, really adds to this story."[62] The final product was marketed as a special, limited edition of Ballentine's Space Blend.

In his role at Open Space Agency, Parr's passion goes well beyond designing glassware for future occupants of space. He is truly a believer in the future of humanity in space, stating, "Our goal is to help create a spacefaring civilization and we are interested in importing all things that we take for granted on Earth and replicating them in the future."

## Real-Life Synthehol

Are there alternatives to alcohol for our spacefaring descendants? One of the best-known alcohol-related products in science fiction is "synthehol" from the *Star Trek* franchise. According to the StarTrek.com website, synthehol is a "generic name for all types of artificial alcoholic beverages whose intoxication effect is all illusion and not chemical." The benefit of this science fiction beverage is that it is supposed to provide the pleasures of alcohol without impairing a drinker's faculties.

Perhaps it should not come as a surprise that there are real-life efforts to create "synthehol-like" beverages. One such project is being conducted in London, England, by Dr. David Nutt, the director of the Neuropsychopharmacology Unit in the Division of Brain Sciences at Imperial College. Dr. Nutt spent ten years "to develop something with all the pros and few of the cons: a synthetic version of alcohol that provides the same warm, lubricating effect in the brain but is not carcinogenic or toxic, cannot make you truly drunk, or give you a hangover—and has no calories to boot."[63] His product is called Alcarelle, which falls under a category of synthetic alcoholic beverages called "alcosynths." Nutt reports that one of the main health benefits of Alcarelle is that it does not produce acetaldehyde, a carcinogen that is a byproduct created by the liver with traditional ethanol-based drinks. Acetaldehyde is connected not only with the unpleasant symptoms of a hangover, but also with numerous forms of cancer and other illnesses.

In a promotional video, Nutt explains, "If you drink (real alcohol), the effect goes up and up and eventually you'll get so drunk you die," but with Alcarelle, "using pharmacological techniques ... you can have an effect that plateaus out—and you can never die."[64] Not dying is certainly a good selling point and Nutt predicts that his product (or similar ones) will replace traditional alcohol by 2050, but prior to that he predicts that "it will be there alongside the scotch and the gin; they'll dispense the alcosynth into your cocktail and then you'll have the pleasure without damaging your liver and your heart."[65]

Like many innovative concepts, Nutt's product has been met with a fair amount of skepticism. There are doubts whether the final product will really be equal to the pleasurable effects of traditional ethanol. Others, regardless of whether Nutt succeeds with his goal, doubt whether it would replace alcohol to any significant extent. Robert Dudley, the author of *The Drunken Monkey: Why We Drink and Abuse Alcohol*, doubts whether this type of product could replace traditional drinks. "My instinct is that we have an intrinsic drive to consume ethanol that can't be replaced.... It's our evolutionary heritage and we just have to deal with it."[66]

Even if David Nutt is successful in his dream to create a safe, synthetic substitute for alcohol, its utilization in space will probably be based less on the health impacts of the product and more on how easily the product can be produced at a future space settlement. If the process is at least as achievable as traditional alcohol, it may be an attractive alcohol substitute to future space settlers.

## Natural Light in Space

Not all alcohol producers are interested in investing in complicated and expensive space experiments that may not pay off for (at best) years. Some drink makers and enthusiasts have found ways to capture the magic of connecting their product with space exploration, but without the great cost or scientific rigor that is usually required for sending an experiment up to the International Space Station or even a zero-gravity flight. One popular option is to utilize lighter-than-air travel. High altitude balloons have provided an opportunity for some companies to provide the appearance that their product is in space and capture some of the publicity and enthusiasm that is currently growing for space exploration.

Probably the first space balloon stunt with alcohol took place on November 18, 2011, when a can of Natural Light beer ascended to roughly 90,000 feet. In fact, this experiment was conducted by a couple of beer enthusiasts rather than the Natural Light brewery. The stunt simply sent cans of that beer to the edge of space, where video images of that product were captured with the Earth in the background. The video proclaimed that their goal was "to make Natty the first beer in space." In the video, they also sent a message to any extraterrestrial beer drinkers who might receive their balloon delivery, with a note that said, "What up aliens! We brought the beer, where's the party?"[67] After reaching its top altitude, it descended back to Earth sixty miles from where it had taken off.

## Ardbeg's "Other" Space Mission

Natural Light was not the only adult beverage to see their product ascend to the edge of space on high altitude balloons.

Even though scotch maker Ardbeg was able to send an experiment all the way to the International Space Station, it was not the only trip to space that whiskey would take. Foster Film Productions has worked with Ardbeg for over ten years on a number of advertising film projects. Inspired by the Ardbeg space mission, they decided to conduct their own mission featuring Ardbeg. They arranged to send a bottle of Ardbeg on a balloon voyage. In a video documenting this stunt, the filmmakers state, "Inspired by the ongoing experiment aboard The International Space Station, two Ardbeg aficionados attempted an experiment of their own."[68] According to filmmaker Ben Foster, the goal was to send a bottle of Ardbeg "further than any bottle of whiskey had ever been in its completed state. Then to bring the bottle to an event and have a showpiece."[69]

Shortly before they launched their balloon from Edwards Air Force Base in California, one of the filmmakers joked, "We don't really know what we're doing, but we've seen Apollo 13."[70] Armed with that level of space exploration

**Ben Foster of Foster Films conducting a high-altitude balloon launch with a bottle of Ardbeg (Fosterfilm Productions LLC; photograph by Matt Mider).**

expertise, they mounted a bottle of Ardbeg Ten and a camera to the balloon in hopes of obtaining crisp images of Ardbeg floating with a dramatic view of Earth below. Their mission succeeded in elevating the bottle to an altitude of over 110,000 feet while collecting stunning video and images of the product floating miles above the ground. After it descended to Earth, they found the balloon over sixty miles away in an area with a prominent warning sign that stated "Danger. No Trespassing. Military Training Area. Unexploded Ordnance Present...."[71] Successfully avoiding unexploded bombs, not only did they find the bottle, but against all odds, they were shocked to find that the bottle of Ardbeg 10 was still intact, stimulating one of them to yell "Holy shit, dude! How did that not blow up?"[72]

As a result of this balloon mission, Foster Films was contacted by Bridgeport Brewing in Maine and arranged for a bottle of their Original IPA to be launched to the edge of space on a balloon in 2018.

## The First "Marsarita"

Mexican tequila company Jose Cuervo also has space aspirations, and decided to conduct their own space stunt to celebrate an "unofficial" holiday. While only dedicated connoisseurs of margaritas would know about the unofficial holiday, National Margarita Day, the Jose Cuervo tequila distillery decided to celebrate this festive day in style when they became the first company to cool a margarita in space on February 22, 2017. Working with JP Aerospace of California, they placed margarita ingredients in a special container that was elevated to the edge of space on a high-altitude balloon. At the top altitude of 100,000 feet, the margarita experiment experienced conditions similar to those on the surface of Mars, or roughly one percent the atmospheric pressure that exists on the surface of Earth. The margarita would also be exposed to elevated ultraviolet and cosmic radiation like on Mars. At a Mars-like (roughly) -92 degrees Fahrenheit, the temperature was more than sufficient to cool the margarita.

As there were no bartenders available to shake this high-altitude beverage, the mix was shaken when the payload was released from the balloon and the parachute was deployed. To add an element of space exploration credibility, they recruited prominent University of Arizona planetary scientist Dr. Peter Smith to assist with the mission and to narrate the promotional video. According to Smith, in addition to pure publicity, "They had seen other products filmed with the earth's curvature in the background and imagined how their products would look from that vantage." Smith added, "I called our drink the 'marsarita' because the conditions are so similar in terms of pressure and temperature."[73]

There have been numerous balloon tests and other space-related stunts over the last few years. This includes the time when London's The Savoy's Beaufort Bar sent one of its new Pop-Up menus on a balloon ride to over 100,000ft to generate some unusual publicity and when Silver Screen Bottlers launched its Ten Forward Vodka to a similar altitude. Stoli Vodka had a different approach; they conducted a zero-gravity flight to mix the first cocktail in weightless conditions. There are many more stories that show the growing interest among alcohol producers, distributors, and others to associate their products with space exploration through these types of stunts.

Over time, it will be interesting to see if high altitude balloons can move beyond the realm of alcohol industry stunts and be used for valuable and innovative research that can benefit the alcohol industry and/or the prospect of alcohol in space.

## Space Technology Benefitting Producers on Earth

### Mars Helps Microbreweries

Most of the projects described thus far have centered on using the space environment to benefit alcoholic products or to figure out how to manufacture booze off-planet. However, there are numerous examples of space technology that were developed for entirely different purposes and now are being put to use to benefit the alcohol industry.

One example is a technology developed for Mars exploration but that is now helping the microbrew beer industry. Robert Zubrin, the founder of Pioneer Energy and the President of The Mars Society, has spent decades investigating and developing methods for "living off the land" on Mars. Zubrin gained a significant level of notoriety for his "Mars Direct" concept that outlined an efficient method to land humans on Mars within a ten-year timeframe. This plan also articulated methods to use the carbon dioxide Martian atmosphere (with hydrogen brought from Earth) to manufacture oxygen, water, and rocket fuel. In the early 1990s, Zubrin built an In-Situ Propellant Plant (a machine designed to use the resources on Mars to manufacture fuel). At the time, Zubrin was employed by the aerospace company Martin Marietta (now part of Lockheed Martin). Funded by NASA, this project was able to manufacture rocket propellant utilizing a simulated Mars atmosphere. The system used a Sabatier chemical reaction that produces methane and water from $CO_2$ and hydrogen. As Zubrin notes in his book *A Case for Mars*, this is a chemical process that "has been in large scale use on Earth for over a century."[74]

How does this benefit the beer industry? Using lessons from his Mars propellant plant, Zubrin and his team developed a method for capturing that $CO_2$ produced during the brewing process. Many breweries, particularly smaller producers, often let the $CO_2$ escape into the atmosphere. However, $CO_2$ is also an important ingredient for beer production. It is what provides the carbonation—the fizz—that consumers expect in their beer. Big breweries have expensive systems to recoup this $CO_2$ so that it can be used later in the manufacturing process, but small microbrewers usually cannot afford this type of equipment. As a result, they usually need to have $CO_2$ "trucked in from supply companies later on, paying an average of $200 to $300 per ton."[75]

Harnessing this technology that was created for Mars exploration, Zubrin and his team created the $CO_2$ Craft Brewery Recovery System, which can collect upwards of five tons of carbon dioxide a month. Zubrin noted, "The intellectual capital being developed in NASA's research and development programs is playing out across the economy, and this is just a small example.... The intellectual capital is the big spinoff."[76] Zubrin estimates that there are several thousand brewers in the United States and many more around the world that could benefit from this product.

## *Spy Satellites for Grapes*

*"For hundreds of years, winegrowers have known that grapes harvested from different areas in their vineyards can produce wines with unique flavors and tastes.... We are now using NASA's advanced remote-sensing technologies to understand the subtle nuances of our vineyards, and with astounding results."*
—Tim Mondavi[77]

Robert Zubrin's innovation was not the first time that space technology has been harnessed to benefit the alcohol industry. For many years, remote sensing has been put to use to help farmers around the world—including the wine industry. In simple terms, remote sensing is observing the ground from the air or in space. Through this technology, farmers can assess the condition of their crops, determine where a specific crop should be planted, and be more precise with irrigation and fertilization, crop forecasting, and many other applications.

Some of the initial remote sensing applications in the wine industry occurred through aircraft-based remote sensing. The wine industry and university partners collaborating with the NASA Ames Research Center created systems with such wine acronyms as CRUSH (Canopy Remote Sensing for Uniformly Segmented Harvest) and GRAPES (Grapevine Remote Sensing Analysis of Phylloxera Early Stress) that used remote sensing devices to prevent the spread of phylloxera (a plant louse highly destructive to grape vines) infestation as well as to observe the health of the grape crops.

Through remote sensing it is possible to define micro-regions within individual vineyards. Daniel Bosch from Robert Mondavi Winery said, "We now identify vine vigor to see weak and strong areas of growth in the vineyard, then we break up how we harvest.... We can taste those differences in the grapes at harvest."[78] The Robert Mondavi Winery was a pioneer in the use of remote sensing. In 1993, they began using remote sensing to detect outbreaks of phylloxera that was ravaging Napa Valley.

Virginia's wine industry has also benefited from NASA technology. NASA Langley Research Center found a way to help their home state's wine industry to better compete with rival wine-producing regions. Through NASA's DEVELOP Program, in which hundreds of students and professionals are given the opportunity to research projects utilizing NASA Earth-observing satellites, a project was instituted to help the wine industry. "In 2015, a handful of participants used data and images obtained by the NASA-built Landsat satellites to help identify the boundaries of vineyards across Virginia as a way to help the Virginia Wine Board, based in Richmond, Virginia, better understand where grapes are grown in the commonwealth."[79]

The Landsat 8 satellite was used to create digitized maps of roughly 250 commercial vineyards. According to Annette Boyd, director of the Virginia Wine Board's marketing office, "If winemakers said, 'I have so many acres,' they wouldn't say 'I have so many acres in this county and so many acres in that county....' We're getting into an additional level of detail with this data that we just didn't have with the voluntary data collection. As we go forward, it'll allow us to really perfect our measurements of Virginia vineyards."[80] It is hoped that the satellite imagery will identify which sites are most suitable for wine growth and which ones are not. This capability will also provide such valuable information as temperatures, soil content and moisture, rainfall, and altitude.

As with the agriculture industry as a whole, remote sensing will play a greater and greater role in the wine industry, as well as the rest of the alcohol industry.

## Mold Reduction in Wine Cellars (and Elsewhere)

Agricultural research conducted on the International Space Station has also helped the alcohol industry. Crop growth experiments on the ISS have provided technology that helps winemakers to keep their cellars healthier for their wine and for their staff members. During growth chamber experiments on the ISS, researchers observed the buildup of ethylene, a plant hormone that occurs naturally. In the proper concentrations, ethylene is essential. It is the substance that causes fruits and vegetables to ripen, but in the confined environment of the growth chambers on the ISS, it accelerates the process and kills the plants.

To solve this problem, NASA partnered with the University of Wisconsin to develop Advanced Astroculture (ADVASC) that was designed to remove ethylene from the chambers. ADVASC also removed bacteria, viruses, and mold that could damage crops.

This system proved to be very successful, and it was soon realized that it could be used for other important applications. As such, the ADVASC technology was the basis for a new air purification system that has become extremely helpful to the wine industry in Napa, California, and elsewhere. While wine cellars are perfect environments for storing wines for long periods of time, these facilities are also ideal environments for mold outbreaks. When a significant level of mold occurs in a wine cellar, its odor can adversely impact the taste of a wine, and the mold can also be harmful to humans if inhaled or ingested. As such, winemakers try to limit or eliminate mold growth in their cellars. The ADVASC system dramatically reduced mold levels in the cellars in which it is being used. According to Andrew Schweiger, from Schweiger Vineyards in St. Helena, California, "Within 24 hours, the amount of airborne mold spores were dramatically reduced."[81] This was not just a random occurrence. Other winemakers noted that within two weeks, well over ninety-nine percent had disappeared from the walls of the cellars. As a result of this success, the air purification system is also now being used in medical facilities, industries, and in many homes. Schweiger reflected, "It's really amazing when you think about all the innovations that are going on up in space, how they can come into a place as unexpected as a winery, which translates to benefits on your dinner table."[82]

## Why Do Alcohol Producers Care About Space?

The alcohol industry has clearly seen the value in partnerships with the space exploration industry, but why is this the case? The most obvious answer is publicity, but based on the stories illustrated earlier, reality is far more interesting and complicated than just advertising campaigns. Nevertheless, advertising and public relations certainly do play a role. Businesses are always looking to find a unique strategy to promote their products. In some sectors of the alcohol industry there is a new urgency to achieve this task, particularly for beer producers.

As beer sales have declined, the commercial space industry has been on the rise, led by technology billionaires who sometimes seem to trailblaze the future by sheer force of will. The alcohol industry may be trying to capture some of that "magic." "Millennials have been particularly captivated by the commercial space industry," said Kellie Gerardi. "I think beverage companies have taken note of that fact, so it makes perfect sense to align themselves with such a sexy, appealing industry."[83]

It is debatable, however, whether being "sexy" would entirely justify the significant expense of space-based experiments and realignment of a brand. If it were just a matter of advertising, all they would need to do is hire the right advertising firm to connect their product with space. Heineken did this in a 2005 television ad depicting a European rover on the surface of Mars. A mission controller declares, "Now, if there is life, the Dutch will find it." The rover then begins to transform into a Heineken beer bar, with a neon Heineken sign up top. The mission controller turns to a young mission specialist, apparently responsible for the transformation, who simply replies, "Now wait." This was a well-produced and successful advertisement, but Heineken did not have go to Mars to produce it. It is doubtful whether anyone believed that this ad transformed Heineken or the beer industry.

Clearly, there is more to this trend than standard advertising or even the "cool" factor. Richard Phillips, a marketing expert and president of Phillips & Company of Austin, Texas, agrees that while the "cool factor" is a contributing factor, it is far more complex than that. "Historically, I believe there are two reasons why they think there is a market. One, geography and the novelty of saying that this was manufactured in space—therefore, it's cool. Two, there is going to be some measured difference in taste, consistency—something they can point to that differentiates them from others. Smart companies understand that 'first to market' is always a good position to be in."[84]

Geography has played an important role in the marketing of spirits, beer, and wine for centuries. For example, the Bordeaux region of France has justifiably garnered a reputation for some of the finest wines in the world. That said, not every vintage from that region is equal. Geography can elevate the status of a lesser quality product that would have less prestige if it were produced a few miles away in another wine region. As such, it is not only quality and taste that people are buying into. "People are also buying into the culture, the expectation that they're part of something bigger," said Phillips. "Building community is the utmost goal of any brand. Having a community of like-minded people, whether they be millennial or the Apollo generation—getting them to commit to a brand and be loyal to it—is about creating shared values and shared experiences."[85] Wine connoisseurs and enthusiasts alike will often build these communities based on personal taste palate and emotional attachment to certain wine regions and they can get competitive in their desire to show that their preferred region is superior to another.

Is space a viable enough destination to carry a brand forward? Fewer than 600 people in human history have been to space, thus it is certainly an exotic and exclusive location—and is associated with adventure and danger. It should be noted, however, that so far no alcohol brand has anything more than a partial connection to space. Alcoholic beverages are not likely to be produced in space in any significant quantity for at least decades.

This being the case, the connection with space alone will not be enough to drive sustainable sales in the current market. "If a company is going to build a brand around space, novelty isn't enough," said Phillips. "People will try it, but they may not come back to it because they'll be loyal to the brands that have built a community for them. Novelty only goes so far, and they'll have to create a brand experience."[86]

Can the alcohol industry successfully build a community and culture around their space aspirations? They may want to look at the success of a non-alcoholic beverage brand that has built a powerful community and loyalty base that transcends their product. Red Bull has successfully marketed their product to be synonymous with excitement and adventure. Red Bull founder and CEO Dietrich Mateschitz explained that company's marketing and culture well when he said, "What Red Bull stands for is that it 'gives you wings'.… It is an invitation as well as a request to be active, performance-oriented, alert, and take challenges. When you work or study, do your best. When you do sports, go for your limits...."[87]

Red Bull has been so successful at building this persona that a culture and community has grown around their brand, regardless of whether all their "citizens" drink or even like their core product. This was by design. When asked whether Red Bull was a beverage company or a media company or something else, Mateschitz replied that it is both. They "communicate 'The World of Red Bull.' Since the beginning it has been a brand philosophy and how we look at the world, rather than pure marketing for consumer goods."[88]

To achieve this unique status, they have gone well beyond words. Red Bull has embraced and supports adventure and achievement—particularly in extreme sports. This includes such activities as the Red Bull Air Races that assemble some of the best aerobatic pilots in the world to compete in an airborne racetrack/obstacle course. They also support racing, surfing, cliff diving, and innumerable extreme sports and teams. One of their most ambitious investments was their sponsorship of Red Bull Stratos that saw Austrian skydiver Felix Baumgartner skydive from the edge of space.

Red Bull has clearly tapped into a concept and a brand that exceeds the value of their core beverage product. It is unclear whether alcohol producers will be able to use space to replicate what Red Bull has achieved, but they may be learning lessons from that power drink powerhouse. Red Bull was able to capture a philosophy of life. Alcoholic beverages might also be able to tap into this approach and tailor a philosophy—a vision for living—for people who are inspired by space exploration. Given the fact that so many alcoholic beverage companies have embraced space in some manner, it would appear that they see real value in that relationship. "Space is seen as attractive for these brands," commented Kellie Gerardi. "They want to attach their brands with this aspirational marketing message. Their beverage and the consumers of their beverage are edgy, they're interesting. They're forward thinking. The fact that there are opportunities and pathways for them to conduct real research on ISS is wonderful for everyone. It's an everyone wins situation—from marketing, science, and the consumers back home."[89]

The current momentum in the field of space drinks is unquestionably impressive, but we still have no idea how long it will be before it is possible and practical to manufacture alcohol and other products in space However, according to Bill Nye, he believes that there will be a market even for the earliest of these products. "Everybody's going to want Martian ice cream and Martian wine and just think of the marketing potential [you] already have.

Like in that ice hotel in Norway, you have this glacial ice in your scotch, that's the thing you do. This will be a thing people do. Historical model, you know that wine that's called port, it would keep longer at sea the same way you make a jam or jelly to preserve the fruit. The port wine was originally created to tolerate long sea voyages, so I can easily imagine a 0G type of spirit that's well-suited to long space voyages."[90]

# 6

# Farming in Space

*"Godspeed little taters, my life depends on you."*
—The Martian[1]

With the extraordinary rise in the number of space liquor projects, one can easily imagine the emergence of distilleries, breweries, and wineries at off-world settlements in the upcoming decades. This may well come to pass, but there are still major obstacles to overcome that may be outside the control of alcohol makers. Most of the materials needed to sustain space exploration do not "grow on trees," but of course, the materials needed for alcohol manufacturing do in fact grow on trees—or at least come from plants.

Producing any significant amount of alcohol away from the safety and resources of Earth will almost certainly require agriculture. In fact, no long-term human habitation is practical without the ability to grow plants.

Food (and booze) are not the only reasons why the ability to grow plants in space is advantageous. Plants can also contribute to life-support systems by producing oxygen, removing carbon dioxide, and by providing a natural way of scrubbing the air, and they also can contribute to the mental health of crew members.

While space agriculture has not been highly publicized in the past, it has recently appeared prominently in popular culture. In the motion picture (and novel), *The Martian*, Mark Watney (played by Matt Damon) grew potatoes on Mars by fertilizing the spuds with his own bodily waste, inspiring some late-night hosts to refer to Matt Damon's "shit potatoes." Nonetheless, as with spirits in space, space agriculture is not science fiction. In fact, it has been studied by researchers and space agencies for decades. *The Martian* (both the book and motion picture) may not have been entirely accurate, but many experts are confident that potatoes and other crops can be grown and consumed on the surface of Mars as well as other off-world locations.

## A Brief History of Space Agriculture

Scientists, futurists, and science fiction writers have speculated about growing food in space for over a century. In his 1880 novel, *Across the Zodiac*, Percy Greg depicts a human voyaging to Mars who utilizes plants to help with recycling waste. A few decades later, the renowned Russian aerospace scientist Konstantin Tsiolkovsky proposed greenhouses to help sustain closed environments in space.

By the mid-twentieth century, however, space agriculture began to move beyond the realm of science fiction authors and futurists. As early as the 1950s and 1960s, studies were conducted for the United States Air Force and the newly formed National Aeronautics and Space Administration that tested the use of algae for oxygen production and the removal of carbon dioxide; and during the 1970s, agriculture experiments were flown to Skylab, including growth tests for rice.

However, as international space ambitions grew, so too did interest in space agriculture. There are currently dozens of projects around the globe that are investigating methods for growing crops in space or are growing plants in simulated Mars or lunar environments. This task is not just a matter of placing seeds in a pot of soil and providing irrigation. According to NASA plant physiologist Raymond Wheeler, "There are different constraints depending on the gravity environment. It's a lot easier if you have gravity in terms of managing water." Wheeler also stated, "I think one of the biggest challenges is still going to be providing sufficient light. On a small scale, you can do this with electric lighting, but as you get larger and larger ... it will take a lot of light. If you do that electrically, your power cost goes up."[2]

One of the biggest unknowns is whether bacteria can survive in the Martian environment. Plant growth on Earth requires innumerable indigenous bacteria to enable healthy growth. A recent study conducted at the University of Edinburgh in Scotland investigated one of the most concerning issues for agriculture (and human health) on Mars. This would be the existence of perchlorates in the Martian soil. The Scottish experiment examined how a bacterium called *Bacillus subtilis* would react to perchlorates. Microbe samples were inserted into a solution of magnesium perchlorate at similar concentrations that would typically be found on Mars. The researchers also exposed the samples to Mars surface-like levels of UV radiation. Under these circumstances, the bacteria did not flourish. In fact, all samples died within 30 seconds. As noted in a *Nature Magazine* article, "the combined effects of at least three components (UV radiation, iron oxide, hydrogen peroxide) of the Martian surface, activated by surface photochemistry, render the present-day surface more uninhabitable than previously thought, and demonstrate the low probability of survival of biological contaminants released from robotic and human exploration missions."[3] Based on these results, researchers were justifiably pessimistic about the future viability of agricultural activities on Mars.

Fortunately, other experts believe that the Mars surface radiation challenge can be overcome. Raymond Wheeler shares this belief. "In general, plants are pretty tough, but it is likely that any crops will already have some intrinsic radiation shielding. If you want humans to access them, you will need some radiation protection." Wheeler added, "I think the plants will probably be fine."[4] Based on recent findings from the Mars Curiosity Rover, the surface radiation on Mars appears to be at a similar level to what exists in low Earth orbit on the International Space Station. Indeed, there have been years of successful experimentation with plant growth on that facility and elsewhere that indicate that radiation on Mars should not be an insurmountable obstacle.

In fact, the perchlorates that were mentioned in the Scottish study may be a more daunting issue than radiation. Not only could perchlorates potentially kill any plant or other organism, but there are also concerns that even if the plants are able to grow successfully, they might have infused some of these toxic chemicals, thus making them toxic to humans.

International Potato Center scientist Jan Kreuze hopes to "try and better understand the genetics and epigenetics of salt tolerance in potatoes, and how we can use it to generate more stress-tolerant varieties. We might also look at soil amendments or treatments to alleviate salt stress by other means."[5]

A preliminary study that appeared in the *International Journal of Astrobiology* highlighted an experiment where two varieties of potatoes were planted in simulated Mars soil. The report stated, "Extreme soil salinity will be an important stressor to the growth of any plants using Martian soil. Under a controlled/protected environment with pressurized atmosphere, a combination of an appropriate sowing method, tolerant genotypes, and soil management will be crucial to achieve yield in such conditions."[6]

Even if these counter-measures for perchlorate are not effective, other options appear to be viable. According to John Connolly, an aerospace professional and owner of Star Creek Vineyard in Texas, "The good thing about perchlorates is that they're highly soluble in water, so we may actually have to wash the perchlorates out of the Martian soil before using it for agriculture."[7] Another option may also be "burning" it; that is, future Martian farmers may have the option of feeding the Martian soil through a furnace to remove the perchlorates before they utilize it for agricultural purposes.

There are also natural methods of reducing perchlorates. Some perchlorates exist in terrestrial soil, but there are types of bacteria that are able to break them down. According to Wieger Wamelink of Wageningen University in the Netherlands, "I would bring those bacteria to Mars and would add them to the soil—let them break down the perchlorate—and that solves most of the problem. I still get chloride, which is still a problem—but not nearly as big a problem as perchlorate."[8] The Dutch team performed experiments with chlorides in soil at about a level of two percent, which is similar to what would be seen on Mars, and they were able to grow some crops that are able to remove chloride from the ground.

There are many other challenges facing the current pioneers of space farming, and some of the biggest questions will not be answered until humanity has established a permanent presence on another planetary body. That said, there has been a remarkable amount of progress in the field of space farming in recent years.

## Potatoes on Mars Project

The home-grown potatoes in *The Martian* may have been a science fiction story, but as it turns out, potatoes may actually be a promising crop for future Martian settlers. To test this hypothesis, an international group of researchers is investigating the feasibility of growing Martian spuds. The group is appropriately called *Potatoes on Mars*, and is a joint project between the NASA Ames Research Center in Mountain View, California, and the International Potato Center (CIP) in Lima, Peru. The project has conducted a series of experiments to determine whether potatoes can grow in a simulated Mars atmosphere and soil.

Julio Valdivia Silva from the NASA Ames Research Center explains, "The initiative came from CIP, with the intention of solving problems around cropping in desert areas as a result of climate change and desertification." Silva added, "Meanwhile, NASA was interested in the project for the need to grow crops in future human colonies outside Earth."[9]

Together, NASA and CIP designed an innovative project to determine whether potatoes could grow on Mars and to find ways to improve potato growth in arid areas on Earth. To simulate the Martian regolith, the team collected soil samples from the Pampas de La Joya desert in Southern Peru. This desert is one of the driest locations on Earth and is considered to be one of the best analogs for Mars. Like Mars, its soil is extremely dry and lacks nutrients that would be abundant in most other places on Earth. It also has "very low levels of organic carbon and the presence of exotic minerals (including salts) and oxidants."[10] Combined, the soil would be considered extremely ill-suited for traditional agriculture.

To create an environment that was more Mars-like, the Potatoes on Mars team utilized a hermetically sealed container roughly the size of a Cubesat (a cubesat is a miniature satellite that is roughly $10 \times 10 \times 10$ cm cubic units each) that contained their unique Mars potato experiment. Inside this container, the experiment provided nutrient-rich water and harsh conditions such as low temperature, low air pressure, and other characteristics that are as close as possible to those on Mars, but still viable for the growth of potatoes. To add a further level of authenticity, the container also simulates light levels for the Martian day and night. Based on the initial findings, Dr. Chris McKay of NASA Ames is optimistic about the prospect of Martian potatoes. "We certainly think that potatoes will grow on Mars if given Earth-like pressure and temperature. It is known that potatoes can grow in cold conditions. The point of our research program is to determine how low the pressure can be and still have growth. We think we can easily get to 1/10 of Earth sea level pressure and perhaps to 1/20. That is still not Martian (1/120) but it's a big step of the way."[11]

McKay believes Martian soil should be suitable for cultivating potatoes because growing potatoes is based less on the nature of the Martian regolith and more on the physical matrix it can become with the help of humans. McKay added, "We would have to add water and any nutrients (potatoes don't need a lot of nutrients)."[12] In fact, many varieties of potatoes are known to be extremely hardy, growing within the Arctic Circle and at altitudes higher than 4,000 meters above sea level. "Wild relatives are found in more extreme habitats, including extremely arid, saline, and frost-prone areas."[13]

The Potatoes on Mars researchers found that a variety of potatoes called "Unique" performed better than other types that they tested. "It's a 'super potato' that resists very high carbon dioxide conditions and temperatures that get to freezing,"[14] commented Silva. Even though the natural conditions on the surface of Mars are far harsher than conditions simulated in this experiment, mission planners and scientists do not intend to grow food crops directly in the Martian environment. All agriculture will take place within artificial environments built by human crews or robots to approximate atmospheric conditions closer to that on Earth. However, finding varieties of potatoes like "Unique" that can grow and thrive in harsher conditions than are common on Earth may not only enable hardier crops, but may also enable crops that do not require as much energy and human labor, or the same atmosphere that humans would require.

**Potato tuber plantings in simulated Mars soil at the International Potato Center in Peru (J. Vandivia-Silva, Bioengineering Dept., UTEC).**

Understanding the lower limits of reliable crop growth on Mars can also have other benefits. This knowledge can influence mission design planning before humans even set foot on the surface. According to Raymond Wheeler, "Understanding the lower limits of survival is also important, especially if you consider pre-deploying some sort of plant growth systems before humans arrive."[15] Wheeler believes that potatoes will be just the starting point. "Wheat and soybeans, things like that, and along with the salad crops, you could provide more of a complete diet."[16]

## *Mars in the Netherlands: Wageningen University*

"Now, if there is life on Mars, the Dutch will find it!" While the Heineken beer commercial mentioned in an earlier chapter was only intended to be a comical piece of advertising, the Dutch may actually help to bring life to Mars. Agricultural researchers at the University of Wageningen in the Netherlands are using simulated Martian and lunar soil to determine whether it will be possible to grow food on Mars and on the Moon. According to biologist

Wieger Wamelink, the project began with a simple question in 2013: "Can you grow crops in these soil simulants?"[17] To answer this question, the university first acquired some of the best simulated Mars and Moon soil available. One of the Mars simulants that the Wageningen team uses originates from volcanic ash from Hawaii and is based on data collected from the Martian surface during the Viking missions in the 1970s. The other simulant is an improved version from the Mojave Desert (Mojave Mars Simulant) that was based on data from the 2008 Mars Phoenix lander that descended to the Martian arctic regions. While both soils approximate some aspects of Martian soil as it is currently understood, neither of these simulants contain perchlorates or account for radiation exposure. The only lunar simulant used by the Wageningen University, called JSC-1, is manufactured from basaltic volcanic ash that originates from the San Francisco volcano near Flagstaff, Arizona.

Within these simulants, Wamelink's team seeded fourteen different plant species during their first experiment, including traditional crops and some wild plants. Since the regolith on the Moon and Mars lack the organic nutrients that are ubiquitous on Earth, they deliberately did not want to inadvertently introduce nutrients by using tap water (which contains nutrients), so they watered the seeds with demineralized water.

As they began the tests, Wamelink was not highly confident of successful germination of the plants. To increase the chances of success, they planted 100 seeds per species with the expectation that only a few of them might germinate. Their crops included radishes, tomatoes, potatoes, garden cress, and arugula. After fifty days, "much to our surprise, most of the seeds did germinate—particularly in the Mars soil simulant. In the Mars simulant, almost all of the seeds germinated, and we ended up with a huge amount of plants—and huge amount of work."[18] The garden cress also did particularly well. It flowered, and in some cases began seeding. Only the potatoes seemed to struggle in this first test, but Wamelink believes that this may not have been the result of the soil itself, but of the conditions in the greenhouse. "In the greenhouse it's always summer, but potatoes need autumn to stop growing."[19]

As for the lunar soil experiments, "it was quite problematic."[20] Most of the seeds germinated, but virtually all of them did not grow well—or they died. Wamelink hypothesized that this could be a result of the fact that the lunar soil is very hydrophobic. In other works, it repels water and may have resulted in the seeds and plants receiving an insufficient amount of water. The lunar soil also contains levels of aluminum, which can be toxic to plants.

Equally concerning, even if crop varieties grew well in the simulated soil, there was a possibility that the plants would not be fit for human consumption. The plants may have been infused with the toxic substances that are contained in the soils, such as heavy metals that are harmful to humans.

During follow-up experiments, Wamelink and his team tested the limits of the soil using remnants from previous harvests to add as many nutrients as possible that exist in Earth soil. They also introduced fertilizer to the soil. As in *The Martian,* their first fertilizer choice was human feces and urine since those will probably be used at actual settlements on the Moon or Mars, but there are strict regulations regarding the use of human waste for agricultural purposes. Instead, a nutrient solution approximating the nutrient levels in human feces was used.

Predictably, this resulted in a significant increase in the harvest. "We had so many green beans, we didn't know what to do with them and the tomatoes were growing so fast that we had to cut them, otherwise they would have grown out of the greenhouse."[21] They also confirmed that the harvested crops were safe to eat. Any heavy metal content was well within acceptable levels for human consumption. This even included the potatoes and carrots that grow in the ground. Once they were proven to be non-toxic, and since this entire project had been supported through a crowdfunding campaign, they invited their supporters to a meal entirely composed of the crops grown in this experiment.

The fact that green beans did so well—they were the best performing crop—has another potential benefit. Only small amounts of nitrogen have been detected in the Martian atmosphere and soil and the Moon has no nitrogen at all. Even the small amount of nitrogen on Mars presents some opportunities, however. Some varieties of plants, such as green beans and peas, are especially efficient at extracting nitrogen (with the help of nitrogen fixing microorganisms) from the atmosphere and injecting it into soil. This could provide at least a partial solution to the nitrogen scarcity in Martian regolith, but if the green bean method proves to be insufficient, human waste could also supply additional nitrogen to the Martian soil.

An efficient flow of water in the soil is vital to irrigate plants properly, but this proved to be a challenge with the highly hydrophobic lunar simulant—and to a lesser extent, the Martian simulant. To address this issue, Wamelink and his team came up with a novel approach—earthworms. Earthworms play an integral role in efficient irrigation within soil, but also in the manufacturing of fertile soil. The earthworms consume dead organic matter such as plants as well as small amounts of soils and then they excrete the mixed material. Their excrement then gets broken down even more by bacteria that releases various nutrients such as potassium, phosphorous, and the ever-important nitrogen.

A major concern was whether earthworms would be able to survive in Martian or lunar regolith. Both types of regolith are much sharper than what worms would ordinarily encounter in Earth soil, particularly lunar regolith, which gets no exposure to the erosion that tends to round the edges of rocks, pebbles, and sand. Since the worms would not survive in the nutrition-free simulant soils, the researchers injected the simulant with pig slurry. It is unlikely that any pigs will reside on the Moon or Mars in the near future, but the slurry would provide the needed nutritive value to the soil, and would also serve as a viable simulant of the human waste that will probably be used on Mars.

Not only did these worms survive, but they appeared to thrive, and even reproduced during the experiment. The experiment began with four worms, and seventy-five days later the researchers were surprised to discover that there were now five worms in the greenhouse. Despite this reproductive success, more detailed tests will be needed to determine the survivability limits of earthworms in lunar and Martian simulants. For example, how do earthworms react to reduced gravity situations? Short of sending worms to Mars or the Moon, experiments need to be conducted in space to simulate these gravity levels, such as with the EU-CROPIS (Euglena and Combined Regenerative Organic-Food Production in Space) mission that will simulate the gravity of both the Moon and Mars (although this satellite does not contain any earthworms). The most relevant earthworm experiment to date was when NASA and international partners sent 4,000 worms to the ISS for six months to determine how they adapted to microgravity. While this was certainly not a perfect analog for the Moon or Mars, the worms performed well in near-weightless conditions, which may bode well for their adaptability in the reduced gravity environments of the Moon and Mars.

Assuming the worms can thrive in these exotic environments, they also present another potential benefit. They could become a dietary supplement for human crews or settlers. "In principal, it's possible. You will be vegetarian for the first few years, but they (earthworms) could become source materials for 3D food printers. It might even become tasty if you mix it with something else."[22]

## Epcot Center

As unlikely as it seems, one of the earliest lunar regolith agriculture programs was conducted by Disney World in Florida. Specifically, the Epcot Center conducted experiments that utilized simulated lunar regolith to grow crops. The lunar simulant had been extracted from a quarry near

Lake Superior that "is the only site that contains 1 billion-year-old, dark gray basalt rocks that are remarkably similar to parts of the Moon."[23] These rocks were then broken down into a more realistic lunar soil by the University of Minnesota.

In a 1986 interview, Bill Eastwood, a scientist at Epcot, highlighted how the lunar-simulated soil would be placed in a greenhouse in an attempt to grow soybeans and wheat. As noted earlier, lunar regolith contains no nutrients, so the Epcot experiment used as few nutrients as possible. Plants like green beans and soybeans are also extremely efficient at extracting nitrogen, courtesy of their microbiome, from the atmosphere and injecting it into the soil. Through rotations of crops, Eastwood hoped that nutrients would build up over time, enabling more sustainable crops.

While many of the recent botanical experiments in simulant tend to be focused on Mars, the Epcot researchers may have been on the right track. Lunar regolith has some advantages over Mars regolith when growing crops. According to noted geomicrobiologist Penelope Boston, the Moon is geologically inactive, and water has not played a role on it for billions of years, "so, a lot of the complicated surface chemistry we have on Mars is not there. The lunar regolith is probably more easily convertible to what we think of as soil—whereas on Mars we have to remediate some of this highly acidic or highly alkaline—or corrosive chemistry, so I think the lunar case might actually be a little bit easier. I can imagine making lunar regolith into soil."[24]

## Crops in Microgravity

### VEGGIE

Gravity will not always be available for space farming. In fact, all agricultural experiments that have occurred in space have been in microgravity—not within the gravitational fields of the Moon or Mars. NASA has been one of the pioneers in advancing microgravity botanical methods. In recent years, they have been operating a plant growth system called Veggie, a small food crop growing system on the ISS. NASA describes Veggie as being "a small vegetable production unit with a passive water delivery system.[25] Designed for space flight, plants can be grown in Veggie using small bags ('pillows') with a wicking surface containing soil and fertilizer."[26]

The first test of this system in space was called Veg-01 that was flown to the ISS in 2014 to conduct initial space-based demonstrations of plant growth in that system. The initial plants selected for this test were red romaine lettuce as well as zinnias. However, these names were not nearly exciting enough for a space mission, so the Veggie team gave them colorful names, such as "Outredgeous" red romaine lettuce and "Profusion" zinnia.

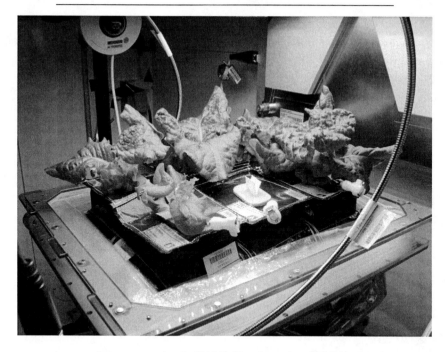

**Lettuce growing in NASA's Veggie project (NASA).**

The initial test faced several challenges, however. The lettuce suffered from inadequate irrigation, while the zinnia suffered from the opposite problem—overirrigation. Nonetheless, even though the overwatering resulted in the zinnia plants being infected by fungal growth and other problems, two of the plants survived and flowered. Samples of both plants were then returned to Earth to be analyzed. Researchers were specifically interested in whether the lettuce was suitable for human consumption, but until the samples were thoroughly analyzed, astronauts and researchers would not be allowed to taste-test them. After a thorough phase of testing, the lettuce produced in the Veggie system was deemed safe for human consumption.

The first in-space taste-test of the space-grown crops generated from Veggie did not take place until 2015, during the time period when astronaut Scott Kelly was conducting a nearly year-long stay on the ISS.

The crops grew for thirty-three days, at which point the surviving plants were harvested. After an extensive sterilization process, the crew consumed half the crop of "Outredgeous" lettuce and the other half was meticulously frozen and packed up to be analyzed back on Earth. In addition to Scott Kelly, other astronauts on board the ISS at the time included NASA astronaut Kjell Limdgren and Japanese astronaut Kimiya Yui. In one of the most unique food reviews on record, the crew gave the space lettuce pretty favorable reviews. Limdgren and Yui said that the lettuce was "awesome." Kelly, on the other hand, stated, "It tastes like arugula." While this description got a lot of mentions in the press, Kelly later revealed that his comment "was sort of a joke."[27]

However, even if it were a joke, there is a scientifically valid reason why the lettuce may well have tasted like arugula. According to plant scientist Gioia Massa of NASA, the plants are subjected to more stress in space than on Earth, which can cause them to be more bitter. This appears to be the result of a defense mechanism of the plants. "Natural protective chemicals made by lettuce, chicory, and related plants, known as sesquiterpene lactones, have a bitter taste."[28] But, having a stronger, more bitter taste may actually be valuable for astronauts. As noted earlier, some astronauts report that their sense of taste dulls after prolonged stays in microgravity, resulting in a preference for stronger tasting or spicier foods. Bitter space-lettuce may well be tailor-made for the needs of astronauts.

As for the zinnias, they are a good test case for longer-term crops, similar to what is needed for tomatoes. Tomatoes are also grown from a flowering plant that needs to flower before it can produce fruit. According to Dr. Massa, "Seeing how crops flower in this system—seeing if there are any limitations—is really important."[29]

Veggie continues to grow various crops in space with the goal of moving the project beyond the status of "experiment" so that it becomes a reliable food source for future crews. According to Dr. Massa, "My hopes are that Veggie will eventually enable the crew to regularly grow and consume fresh vegetables."[30] The nourishment of the crew is vitally important, but there are other benefits as well, including the psychological well-being of astronauts in long space missions. As Scott Kelly noted, "It's kind of nice to have some flowers up here. You don't see much that is alive and growing besides the six of us here."[31]

## LADA

Veggie is not the only botanical endeavor on the International Space Station. In the Russian Zvezda module at the ISS is a greenhouse called Lada—named for an ancient Slavic goddess of spring and fertility. Lada is "a small, low-cost growth chamber, to research plant development and test methods of growing plants aboard the ISS."[32] It was developed using principles learned from a larger unit called "Svet" that had been housed on the Mir space station. Svet had studied wheat and other crops.

Lada began operation in 2002 to determine whether plants grown in space are safe for human consumption, to examine the interaction of microorganisms on the plants in the space environment, to test sanitation processes for the harvested plants, and to learn how best to efficiently grow crops with limited resources. Mizuna, a leafy plant indigenous to Asia, was the first crop grown in Lada, but other crops that have grown in this unit include peas, tomatoes, radishes, and wheat.

"Growing food to supplement and minimize the food that must be carried to space will be increasingly important on long-duration missions," noted Shane Topham, an engineer with Space Dynamics Laboratory at Utah State University. "We also are learning about the psychological benefits of growing plants in space—something that will become more important as crews travel farther from Earth."[33]

One of the most surprising findings from Lada is that overwatering crops may actually be beneficial to crops in microgravity. They found that crops sprouted and grew twice as fast. According to Topham, "This suggests the conservative water level we have been using for all our previous experiments may be below optimal for plant growth in microgravity."[34]

## MELiSSA

The European Space Agency (ESA) is taking the holistic approach to the challenge of space farming. Since the late 1980s they have been pursuing a concept that integrates agriculture and life support that is loosely based on the process that occurs in an "aquatic lake," where the metabolism of algae and plants are able to process waste products. This effort is called the MELiSSA Project (Micro-Ecological Life Support System Alternative). MELiSSA was instituted in 1987 by a French engineer named Claude Chipaux with the goal of studying crops in atmospherically closed chambers. According to Christophe Lasseur, Chipaux "had the will to see European participation for very long-term missions."[35] Initial work on this project was largely focused on using microbes for waste management, but the project eventually shifted to focusing on plants and how to create a full closed system that continuously processes human waste and breaks it down to help fertilize and water crops. "The driving element of MELiSSA is the recovering of food, water and oxygen from organic waste, carbon dioxide and minerals, using light as a source of energy to promote biological photosynthesis."[36]

Chipaux and his team started modestly, launching two microbial strains on a Chinese Long March rocket in August of 1987. Based on the success of the initial experiments, the European Space Agency (ESA) took over the MELiSSA project in 1989.

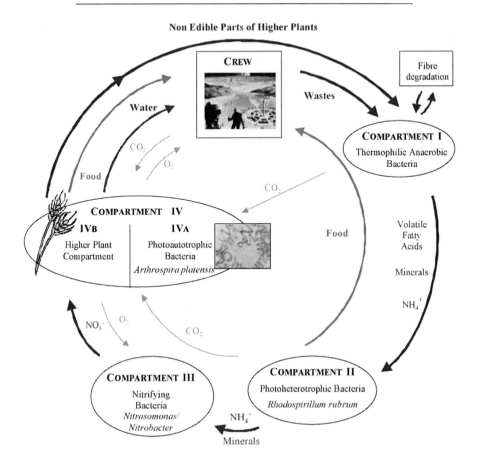

Non Edible Parts of Higher Plants

CREW

Water

Fibre degradation

Wastes

COMPARTMENT I
Thermophilic Anaerobic Bacteria

$CO_2$

$O_2$

$CO_2$

Food

COMPARTMENT IV

IVB
Higher Plant Compartment

IVA
Photoautotrophic Bacteria
*Arthrospira platensis*

Food

Volatile Fatty Acids

Minerals

$NH_4^+$

$NO_3^-$

$O_2$

$CO_2$

COMPARTMENT III
Nitrifying Bacteria
*Nitrosomonas/ Nitrobacter*

$NH_4^+$

Minerals

COMPARTMENT II
Photoheterotrophic Bacteria
*Rhodospirillum rubrum*

**A chart detailing the concept of the MELiSSA Project (MELiSSA Project).**

MELiSSA has grown since its founding and now involves dozens of organizations—mostly universities—in both Europe and Canada. Since their goal is quite ambitious and crosses many disciplines, they initiated a step-by-step phased approach to achieving their goal. MELiSSA is broken into five phases—most of which are proceeding concurrently: (1) Research and development, (2) Preliminary space flights, (3) Ground and space experiments, (4) Terrestrial and commercial applications, and (5) Education and communication. However, this overview of MELiSSA will only highlight activities in phases 2–4.

Thus far the MELiSSA Project has flown ten experiments in space. The most recent was launched to the ISS in December 2017. This experiment, called the Artemis Pilot Project, investigated how photosynthesis takes place in space. According to an ESA statement, "They loaded the microalgae *Arthrospira*, commonly known as spirulina, into a photobioreactor, a kind of cylinder bathed in light. On the Space Station, carbon dioxide will be transformed by photosynthesis into oxygen and edible biomass such as proteins."[37] Spirulina can be used to produce oxygen in space habitats as well as remove $CO_2$ from the atmosphere. One of the questions that MELiSSA scientists formulated was whether this bacterium would be able to function as effectively in space while being subjected to space radiation and microgravity.

In addition to broader contributions to the biological system they hope to create, spirulina also is highly nutritious and can be used to feed the crews. In fact, the DEMES experiment (DEmonstration of MELiSSA Snacks) sent cereal bars made with Spirulina to the ISS. Italian Astronaut Samantha Cristoforetti was the first person to eat food containing spirulina on that orbiting laboratory.

As mentioned above, phase three of MELiSSA focuses on ground and space experiments. The most substantial portion of this phase is the MELiSSA Pilot Plant that was built in 2009 at the University Autònoma of Barcelona in Spain. This research facility "hosts a multi-compartment loop that is extremely airtight, like the ISS," explained Lasseur. "Last year, we ran a photobioreactor with a culture of algae that has succeeded in keeping 'crews' of three rats alive and comfortable for almost six months at a time. The algae produce the oxygen and trap the $CO_2$, the rats do exactly the reverse."[38] While the "rat crew" system is much smaller than will be required for humans, it nevertheless provided extremely valuable knowledge that will help for future systems to support human life.

The MELiSSA technology is also being put to practical use, providing a water recycling system at the Concordia research station in the Antarctic that is co-operated by France and Italy. While drinking water is obtained by melting snow, water that is used for cooking and washing, which is known as "grey water," is recycled using a MELiSSA-derived system.

This type of technology utilization is one of the primary goals of the fourth phase of MELiSSA. This phase identifies technologies or techniques that have benefits for terrestrial use and/or have commercial applications. There have been numerous spin-off applications in water recycling, biofuel, biotech, and other areas. In addition, four spinoff companies have already been created as a result of this program that are actively using technology and techniques from MELiSSA in the private sector.

In fact, some of these techniques have already helped the alcohol industry. To improve their water conservation while brewing beer, the monks at La Trappe Abbey's brewery, The Koningshoeven brewery, in France utilized

MELiSSA Project technology. According to an ESA online article, "To make their beer-brewing process more sustainable, they chose the suite of techniques developed for spaceflight to renovate their factory and recycle more water. Upon its completion in 2019, the system should reduce the amount of water used to brew their beer by 80%."[39] Christophe Lasseur praised this terrestrial application, stating, "In this case we applied our expertise and technological developments in a factory, but they could just as easily be incorporated into a hotel or other operation."[40]

While these phases have generally moved in parallel, MELiSSA cannot advance with many of their goals until some major decisions are made by world space agencies. There is still disagreement on what the next step for human space flight should be for the international partners. Will the coalition of nations choose to go back to the Moon, will they build a cis-lunar facility (like the proposed Deep Space Gateway), or will the partnership decide to go directly to Mars? There is overlap in many of the systems used at the various potential destinations, but there are also considerable differences. As such, MELiSSA and ESA leaders are hesitant to advance some of their more ambitious goals until clear decisions have been made regarding the future of human space exploration.

Nonetheless, looking forward, scale may be the biggest challenge for the future. The earthbound MELiSSA facility is currently too large to go into space with our current capabilities, but the MELiSSA project managers are

**Artist's depiction of a future greenhouse on Mars (Versteeg/Spacehabs).**

confident that the system they are developing will be able to be adapted in the future. "MELiSSA is very scalable," according to Christophe Lasseur. "All the mathematical modeling, testing, validation ... all of this is part of MELiSSA."[41] Much of this work may need to wait until international policy leaders decide on the elusive destination(s) in space.

Regardless of the destination, Lasseur is optimistic regarding the future of the MELiSSA Project. "We are still working on the complete loop—from waste to food, oxygen, and water. I think in ten years, we should be quite advanced."[42]

If Lasseur is correct, this would be a major step in enabling a sustainable human presence is space. As former Director General of ESA Jean-Jacques Dordain stated, "The validation of MELiSSA's highly regenerative life support processes is a mandatory step towards future long-duration human space exploration missions."[43]

## Mars Analog Bases: Simulating Mars

As humanity prepares to voyage into deep space, researchers around the world have built facilities designed to simulate conditions on other worlds, including testing concepts for agriculture at these sites. Some of these analogs are placed in locations that have similar qualities to the planetary body they are simulating; others create closed habitats that are only intended to approximate life inside a space vehicle or ground facility; and there are some that just focus on simulated soil for plant growth.

Haughton Crater on Devon Island in the Canadian Arctic is one of the best-known Mars analog locations in the world. Devon Island is the largest uninhabited island in the world and is home to Mars analog field studies every year. In partnership with the Mars Institute, NASA, and the SETI Institute, the Haughton-Mars Project (HMP) was established by planetary scientist Pascal Lee. According to Lee, "This project uses the polar desert setting, Mars analog geology, and harsh climate of the High Arctic to test technologies and strategies for the future human exploration of both the Moon and Mars."[44]

Every summer, researchers converge on this site to conduct science (geology and microbiology) field tests and other valuable research that will enable humanity to go to Mars. The HMP Research Station is home to the Arthur Clarke Mars Greenhouse (ACMG) that was donated to HMP by SpaceRef in 2002. According to the SpaceRef press release, "ACMG will support scientific and operations research in an operational setting that is relevant in unique ways to Mars—each at a specific level of fidelity and complexity....

Ultimately, through a sequential and iterative program of experimentation, it is hoped that a better understanding of the operational challenges faced by future astronauts on the surface of Mars (or other planetary bodies) will be gained."[45] Between 2006 and 2011, in partnership with the Canadian Space Agency and the University of Guelph, the ACMG was able to achieve the goal of a year-long autonomous operation. According to Pascal Lee, "Once set up in one summer, [ACMG] could be left unattended for an entire year, and produced a new round of crops for the following summer by the time we returned to the site to open up camp." Lee added, "Crops of the first summer died out in late Fall, as planned (long after we had left camp), but a new plant growth tray was activated autonomously come spring, and a new harvest was available by summer. Although plants were not growing throughout the winter (that was never the goal, as the sky gets permanently dark and the greenhouse no longer acts as a greenhouse), all electronic systems were nevertheless kept warm and operational via a combination of solar and wind power."[46]

Some of the crops grown at ACMG include Arabidopsis (a hardy, small flowering plant related to cabbage and mustard) and lettuce. Lee also envisions fast-growing crops such as tomatoes to be well suited for future Mars-based farms.

ACMG showed that an autonomous greenhouse that will continue maintaining crops between Mars crews is certainly viable, but will also be challenging. On Earth, the agricultural industry has devised ways to autonomously harvest crops as well. With enough resources, this could be possible on Mars. However, since plants are extremely fragile and these systems will be challenging to perfect, Lee noted, "I would be reluctant (and would actually object to) having crews on early missions to Mars rely on any plants they would have to grow themselves, or via autonomous systems. In my view, any plant growing on early missions would be in the 'nice to have' category, but should not be part of the critical path to mission success until a substantial infrastructure and permanent human presence is established."[47]

## BIOSPHERE 2

Perhaps the best-known space analog facility is Biosphere 2. Located near Tucson, Arizona, construction of Biosphere 2 began in 1986. The goal of this facility was to conduct research and develop plans to enable self-sufficient space colonies in the future.

This facility includes over 300,000 square feet of mostly enclosed ecosystems, including an ocean with a coral reef, mangrove wetlands (tropical coastal wetlands), tropical rainforest, savanna grassland, and a fog desert (a desert where fog provides most of the moisture). Raymond Wheeler described Biosphere 2 as "larger than what most space agencies might envision for early space missions, but their goals of understanding closed ecological systems and bioregenerative approaches for human life support provided insights into the challenges for agricultural and biological approaches for space life-support systems."[48]

While the facility gained notoriety for not achieving its goal of compete self-sustainability, Biosphere 2 was still able to provide 80 percent of the food for eight crew members over a two-year mission. This project got heavily scrutinized for the fact that it ended up having to pump oxygen into the facility only eighteen months into the mission, but they gained many valuable insights for future planners of settlements on Mars and elsewhere.

In fact, the issues that plagued that mission do not seem to be the largest challenges for space farming according to Taber MacCallum, one of the original Biosphere 2 crew members. He views the biggest challenge as "creating that area of pressurized volume so that you can grow crops ... the next biggest challenge is the power and thermal aspect of getting that much energy into your base or station and the heat back out." MacCallum added, "As for ECLSS (Environmental Control and Life Support Systems), the crops are supporting ECLSS, so it takes a load off the life support system. Crops are great at removing things like ammonia and cleaning the air and cleaning water and obviously removing $CO_2$ and making oxygen."[49] However, MacCallum also noted that while plant growth makes life support systems easier, it adds some challenges, such as the amount of power consumption that is needed to maintain them, as well as the pressurized volume that needs to be constructed and maintained to sustain these plants.

Biosphere 2 received a great deal of criticism at the time for not achieving complete self-sufficiency. In reality, it was probably naive to expect such an ambitious experiment to work perfectly; particularly on the first attempt. When one looks at the knowledge gained from this simulation and the lessons learned for future space settlements, however, it should be considered a success.

## Alcohol at Biosphere 2

While the majority of the original Biosphere 2 mission was taken up with their official duties, crew members did eventually succumb to temptation and attempt to make some homemade alcohol. According to Mac-Callum, "We tried to make banana wine, which for some reason didn't ferment very well. We also tried a wheat sour mash—and it certainly was sour. We did try to produce some alcohol, but none of us found ourselves rolling drunk on the floor—but it was a nice break to have a couple of drinks—as un-tasty as they were."[50]

## THE UNITED ARAB EMIRATES

Not all analog sites are in North America. The United Arab Emirates (UAE) have become passionate about Mars exploration and are now pursuing some ambitious Mars simulations. In 2017, the Mohammed bin Rashed Space Center (MBRSC) in Dubai announced plans aimed at solidifying that nation's role in the future of Mars exploration. The Mars 2117 Project was launched with the goal of building a city on Mars by the year 2117. However, this project also includes some near-term goals. The first major program associated with the 2117 initiative is called Mars Science City (MSC). MSC will be a simulated Martian city built in the outskirts of Dubai and will consist of 1.9 million square feet (approximately 176,500 square meters) of enclosed facilities. According to the press release announcing the facility, "This project will include advanced laboratories that simulate the red planet's terrain and harsh environment through advanced 3D printing technology and heat and radiation insulation. It seeks to attract the best scientific minds from around the world in a collaborative contribution in the UAE to human development and the improvement of life. It also seeks to address global challenges such as food, water and energy security on earth."[51]

In this same press release, Sheikh Mohammed bin Rashid said, "The UAE seeks to establish international efforts to develop technologies that benefit humankind, and that establish the foundation of a better future for more generations to come. We also want to consolidate the passion for leadership in science in the UAE, contributing to improving life on earth and to developing innovative solutions to many of our global challenges."[52]

Agriculture will play an important role at MSC and enable the UAE to expand agricultural research started at UAE universities to investigate suitable crops for Mars. According to Rashid Al Zaadi of the UAE Space Agency, some of the reasons they are interested in agriculture and why the UAE is well positioned to contribute to this growing field are that "there are similarities between Mars and the desert." He then added, "When we get there, we'll have to eat."[53] Specifically, they are considering space-related research on the date palm tree. This tree is indigenous to the Middle East and thus grows in extremely arid areas—and does not require a lot of nutrients in the soil.

As plans for MSC move forward, planners in the UAE intend to open up this facility to the international community. "Working with collaborators internationally, researchers hope to build on research already done on space agriculture involving plants such as lettuce, strawberries, and tomatoes—three foods already shown to have thrived when grown in space."[54]

## CHINA

China also has ambitious plans for space exploration in the upcoming decades, including the space agriculture arena. In 2013, China began construction on Yuegong-1 or Lunar Palace-1 at Beihang University, and it was officially unveiled in January 2014. According to Yuegong-1 chief designer Liu Hong, the 1,600-square-foot facility contains a bioregenerative life support system (BLSS), where "humans, animals, plants and microorganisms co-exist in a closed environment, simulating a lunar base. Oxygen, water and food are recycled within the BLSS, creating an Earth-like environment."[55] The goal of this facility is to prepare for future stays on the Moon, Mars, or other planetary bodies. In May 2017 they began a 370-day series of tests, during which eight students took turns living in the facility.

At the completion of the mission, Liu claimed, "We have recycled 100 percent of oxygen and water that human beings need and 80 percent of food. The system is 98 percent self-sufficient. That's to say if we need 100 kg of supplies, we can regenerate 98 kg in the system by recycling."[56]

China has also not restricted their space agriculture efforts to the ground. Chinese astronauts (or taikonauts) successfully grew crops on their Tiangong-2 space station, including rice as well as an edible weed called thale cress. The purpose of the experiment was to examine how microgravity conditions impact the flowering and growth rhythm of plants. In an interview with the Chinese news service Xinhue, chief scientist for plant research on Tiangong-2 Zheng Huiqiong stated, "So far the plants on Tiangong-2 have been growing well. Some *Arabidopsis thaliana* are blooming, and the rice is about 10 centimeters tall."[57]

Most of the Chinese agriculture efforts—in space and on Earth—have been quite similar to ones conducted by other nations over the past few decades. However, China recently undertook a truly unique space plant experiment. In January 2019, the Chang'e 4 (named for the Chinese goddess of the Moon) spacecraft landed on the far side of the Moon. This lunar lander carried a small aluminum biosphere (approximately 3 kilograms) that contained potatoes, cotton, rapeseed, *Arabidopsis* seeds, and yeast, as well as a small number of silkworm eggs. "Combined with air, water, and a special nutrient solution, the container constitutes its own complete ecosystem, with the potato and *Arabidopsis* breathing out oxygen after taking in the carbon dioxide exhaled by the silkworms."[58] Not only will this be the first time such an experiment is performed on the Moon, but the Chinese could also gain valuable data on how well plants grow in a reduced-gravity environment. In fact, the mission did see initial germination of the seeds, but they were quickly killed by the long lunar night. Nonetheless, they showed that plants can grow in lunar gravity.

Researchers in China are working on multiple other space agriculture projects, and as their space ambitions move forward, it is almost certain that their status as a world leader in space-based agricultural methods and technologies will also grow.

## Reduced Gravity

Despite the existence of so many projects around the globe investigating the prospect of growing food crops in microgravity or in simulated Moon or Mars regolith, there still remains a major variable that researchers have not been able to simulate in a sustained manner. This would be the reduced gravity levels on both the Moon and Mars. Gravity on Mars is one-third what it is on Earth and the Moon has only one-sixth of Earth's gravity. Periods of reduced levels of gravity can be simulated in parabolic flight (such as on the "Vomit Comet" airplane) as well as in the use of clinostats (a device that rotates to expose plants to different levels of gravity). However, sustained simulation of these levels of gravity or at a scale that would enable a better understanding of the impact on human physiology or in plant growth have been limited.

Short of sending missions to the surface of the Moon or Mars, are there any viable options to simulate these gravity levels? For over a century, researchers, futurists, and science fiction authors have advocated the use of artificial gravity by harnessing centrifugal force. In Robert Zubrin's *The Case for Mars*, he advocates producing an artificial gravity system using a tether in which two objects (one being the crew vehicle) rotate around a center of gravity. Using such a system would greatly reduce the deteriorating physical effects of long-term weightlessness that crew members would otherwise have to endure. To put this concept to the test, in 2001 Zubrin and The Mars Society announced a project to create a rotating satellite called Translife that would study the impact of Mars gravity on mice. The mice would be allowed to reproduce within the Mars gravity and researchers would be able to observe for the first time how Mars gravity impacted reproduction in mammals, as well as if Mars gravity would be "an effective countermeasure for mammals against the physiological deterioration that accompanies long-duration space-flight in zero gravity."[59] Initially, this project was going to be conducted in partnership with Elon Musk (who went on to found SpaceX), but the project was never able to achieve its goal of sending this experiment into space.

Other concepts have also been proposed over the past several decades. Arguably the best known of these was advanced by Princeton physicist Gerard K. O'Neill. O'Neill's concept proposed the construction of cylindrical-shaped space stations that would spin to create gravity for its occupants. Unfortunately, very little real work has been conducted to move forward on the concepts outlined by Zubrin, O'Neill, and others advocating artificial gravity.

## Eu CROPIS

To help shed new light on the mystery of artificial gravity, the Eu:CROPIS (Euglena and Combined Regenerative Organic-Food Production in Space) satellite was launched by SpaceX in December 2018. Originally proposed by DLR (German Aerospace Center) Institute of Aerospace Medicine in Cologne and the Cell Biology Division of the University of Erlangen, Eu:CROPIS will simulate both Lunar and Martian gravity. The satellite was sent to an Earth orbit of 370 miles (600 kilometers) carrying two greenhouses that will germinate tomato seeds and be monitored by sixteen cameras. When the first

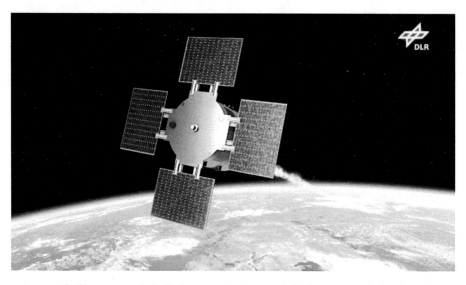

An artist's illustration of the Eu:CROPIS mission launched into space in 2018. This satellite will investigate how tomatoes grow in simulated lunar and Mars gravity (DLR).

greenhouse is activated it will begin to rotate, simulating gravity conditions on the Moon (0.16 that of Earth gravity) for six months. The satellite will then adjust its spin rate to replicate Mars gravity (0.38 that of Earth gravity) and activate the second greenhouse and spin at this speed for another six months. By the time the second greenhouse is activated, the seeds will have been exposed to a similar amount of cosmic radiation that would occur on a six-month voyage to Mars. "Ultimately, we are simulating and testing greenhouses that could be assembled inside a lunar or Martian habitat to provide the crew with a local source of fresh food. The system would do this by managing the controlled conversion of waste into fertilizer,"[60] said DLR biologist Jens Hauslage.

The experiment will use a "trickle filter" that will use synthetic urine to help fertilize the tomatoes. The system will also use *Euglena*, a type of photosynthetic single-celled organism, to help produce oxygen and protect against excess ammonia, and a day/night cycle will also be created utilizing LED lighting. If successful, Eu:CROPIS will help fill a large void in our understanding of how diminished gravity impacts biological systems such as human bodies and plants. Since plants have been shown to grow and thrive in microgravity, most experts believe that they will grow in Moon or Mars gravity, but until this assumption is put to the test, we will not know for certain.

## Terroir, Mars-oir, Lun-oir, or Non-oir

Let us assume that humanity will be able to grow safe food on the Moon and Mars. That being the case, how will it taste? Will these alien locations inject their exotic personality into the flavor of the crops? On Earth, the impact of geology and soil has been of particular interest to producers of alcohol for centuries. Most professionals in the wine industry will speak of specific characteristics from a wine that reflect the geography and soil characteristics in which the grapes grew. This is called "terroir." According to *Merriam-Webster Dictionary* (online), terroir is "the combination of factors including soil, climate, and sunlight that give wine grapes their distinctive character."

For an even deeper examination of what terroir is, Penelope Boston explains, "Plants pick up a great deal of extraneous materials from their environment. Sometimes those factors are metal (irons). Sometimes those are organic materials—and there is also the effect of the seasonal cycles. So, you put all that stuff together ... and produce this unique flavor that is only produced by the precise set of ingredients."[61]

Terroir is big business in the wine industry. Even most non-wine enthusiasts have heard experts talk about the unique attributes of wines from Bordeaux, France, or Napa, California, or the Barossa Valley in Australia. These geographical designations can play a large role in the popularity and cost of specific wines, regardless of their actual objective quality or characteristics.

Within the wine making industry, there tends to be agreement that terroir does impact the taste of their products. Vineyard owner and aerospace professional John Connolly does not think that there is any doubt on this question. He stated, "I know sometimes people roll their eyes at the wine experts, but you can definitely taste the difference between a south Australian shiraz and a Syrah grown somewhere else, even though it's the same grape. I believe there are people who can literally taste the terroir of where that grape has come from. It seems impossible to plant something like a wine grape and not have the soil it's planted in have an impact on the taste."[62]

The wine industry is not alone in understanding the commercial value of terroir. Many agricultural products have also used "terroir" to advertise the unique flavors and characteristics of their products. Penelope Boston explains, "I've heard it most recently applied to Sake. Chocolate has terroir. Coffee has terroir. Chilis have terroir.... In New Mexico where I live, we're famous for chilis. Chilis from the Hatch Valley in New Mexico are famous because they have their terroir and nobody else can call their chilis 'Hatch Chilis.'"[63]

It is easy to understand how wine would be particularly susceptible to local geographical influences. The taste of wine is heavily dependent on the grapes from which it is made. On the other hand, beer contains several ingredients, and spirits go through a distilling process that could potentially remove indigenous characteristics. For example, makers of whiskey—particularly scotch whiskey—have often boasted about the superior "terroir" of Scotland or other distilling locations. However, in recent years, there has been skepticism expressed within the whiskey industry regarding to what extent terroir actually influences the final taste of the product. David Blackmore, the brand ambassador for Ardbeg scotch whiskey, believes that "most of the talk in the scotch industry about terroir is bullshit."[64] While "scotch whiskey" must be produced in Scotland, the ingredients do not need to come from that country. Scotland boasts numerous fields of barley, but a large percentage of the barley that eventually is utilized in whiskey comes from other countries. According to Blackmore, "It's not like a wine. We are distilling it. We're doing that extra process which would take a lot of that terroir out of it. The generation of ambassadors in the Scotch industry before me all spouted that the water (for some Scottish distilleries) was somehow unique—

and gave the character to the distillery. There may be a tiny grain of truth there somewhere, but the science has not been done, so they can't claim it."[65]

The debate about how extensively terroir impacts distilled alcohol or other alcoholic products continues, but few dispute the fact that terroir can impact food and drinks (including wine). Even if Mars or lunar crops are toxin-free, these environments could very well have an impact on the taste of these products.

> Terminology: If humanity can settle Mars in future years and create an indigenous wine industry, how will Terroir or "Mars-oir" impact the taste of these products? Will an entire new terminology need to be established? Martian soil lacks the nutrients and microbes that Earth soils have in abundance, and its chemistry differs in many other regards. The soil has large quantities of iron oxide and sulfur and it contains various salts such as perchlorates that are potentially toxic to humans. If grape vines are able to grow from the Martian soil in a contained environment, how much of the chemistry and characteristics of that soil will be taken in by the plants? When Martian drinkers are swirling their glasses (in ⅓ gravity), sniffing and tasting, will some of the Mars descriptive—terroir—words be:
> "Rusty"
> "Salty"
> "Sulfury"
> "Metallic"
> "Bitter"
> "Astringent"

## Plants as Friends?

Nobody will argue that the primary goal of advancing agriculture in space is clearly for food production and other tangible products to keep crews (or settlers) alive and healthy. However, as referred to earlier, there is a growing area of interest regarding the psychological impact of plants on humans. Raymond Wheeler asks, "Is it part of the human condition to feel better and more physiologically well when you have plants around you? We have evolved on a planet with plants around us and we eat plant foods all the time, so if you remove humans from that type of environment for a long time, how does that affect them?"[66] In fact, the noted Harvard Biologist E.O. Wilson has written extensively about this topic—or what is called biophilia, which means "love of life"—or the "innate tendency to focus on living things, as opposed to the inanimate." Wilson argues that it is a vital part of humanity that does not receive the attention that it should. According to Wilson, "The biophilic tendency is nevertheless so clearly evinced in daily life and widely distributed as to deserve serious attention."[67]

Plants are required for long-term and sustainable human activities on other planets, but the role of biophilia should be embraced as a secondary benefit to human space exploration. Mission planners have always understood the value of trying to create a "home away from home," and this concept could play an important role in achieving that and maintaining the psychological health of crew members. If a botanical system can be created that simultaneously feeds the crew, contributes to the life support systems, and improves the psychological well-being of the crew, it would clearly be advantageous to do so—particularly if the same system could also be the basis of space-based alcohol production.

## Biggest Challenges

As highlighted earlier, space agriculture has been making great strides in recent years, but as the prospect of actual space settlements becomes more realistic, there still remain some major challenges to overcome. The ability to grow a small amount of lettuce on the ISS is promising, but current space-based growth experiments are nowhere near the size or efficiency required to maintain a permanent human presence in space. As such, scale may be the biggest challenge for the future of space agriculture. It is unlikely that humanity will have the ability to erect a structure like Biosphere 2 on other worlds anytime soon. That said, great progress is being made that confirms that it is possible to grow crops in microgravity, and is most likely possible in the regolith of other planetary bodies. Yet it remains unclear how we will scale these capabilities up to a level sufficient to feed even a small outpost on the Moon or Mars.

According to Raymond Wheeler, on Earth "each of us requires between sixty and eighty square meters of plant production for all of our food production. We're decades away [from] developing that scale of agriculture."[68] To feed even a small permanent outpost of twenty to thirty people will require a major allotment of space utilizing traditional methods, but even with more efficient hydroponic and aeroponic methods, it would still take a significant volume of space to sustainably feed crews of any size.

Power is also a significant challenge. All destinations in space that humanity is likely to visit in the foreseeable future have access to sunlight, but the exposure to that sunlight may not always be optimal for reliable crop growth for reasons that include seasonal variation in light, distance from the

Sun, potential dust storms (on Mars), and in the case of the Moon, an especially long day/night cycle. Special lighting and environmental controls will be needed to assure that crops receive at least the optimal level of sunlight and atmosphere. This is essential for healthy and sustainable agricultural pursuits in space. There are viable options, but they also require a significant amount of power. Wheeler noted, "If you use concentrators to collect light and then convey it with fiber optics or light pipes in your protected environment, you can spread it out to whatever space you want."[69]

## Synthetic Biology

> "Would it be possible to genetically engineer a plant or breed a plant that could not only tolerate that high ultraviolet [radiation], but exploit that higher energy and be more productive in a terrestrial planet and then the plants in question would be hops and barley?"
>
> —Bill Nye[70]

Most space agriculture experts agree that volume of space and power required are some of the most formidable challenges that will need to be overcome if there is any hope of growing enough crops to feed a space settlement or colony. What if there was an alternative that did not require as much space, and could not only help generate food, but also produce fuel and other essential materials? Synthetic biology might help provide this solution. The White House National Bioeconomy Blueprint defines synthetic biology as "the design and wholesale construction of new biological parts and systems, and the redesign of existing, natural biological systems for tailored purposes, integrating engineering and computer assisted design approaches with biological research."[71] Dr. Lynn Rothschild from the NASA Ames Research Center expands this definition even further, stating, "Synthetic biology enhances and expands life's evolved repertoire. Using organisms as feedstock, additive manufacturing through bioprinting will make possible the dream of making bespoke tools, food, smart fabrics and even replacement organs on demand."[72] If successful, this technology could also operate at modest power levels, within a limited space, and at room temperature.

Explorers could theoretically bring along a variety of feedstock cells that represent a negligible amount of mass and space, allowing those cells to replicate on their own to create innumerable products. Rothschild gives the example of harvesting latex for rubber. "Why not engineer a yeast cell to produce the rubber hydrocarbons from carbohydrates directly?"[73] This technology is not science fiction. Tire manufacturing companies such as Goodyear and Michelin have partnered with several biotech companies to utilize this process to improve their products. Other companies are using synthetic biology techniques to create products made of spider silk that are extremely strong and flexible. There is already work underway to utilize this technology for food products, including growing meat in a lab. According to *Wired*, "Scientists have been culturing meat in labs for years, but Just and other startups like Finless Foods, which is growing fish meat, have been feverishly pursuing this so-called 'clean meat' of late."[74] With this approach, all you would need would be a few meat cells to grow a large number of meals. "Theoretically from one little piece of meat you can create an unlimited amount,"[75] stated Mike Selden, CEO of Finless Foods.

Over the past several decades there have been debates regarding whether Martian settlers will be vegetarian or vegan. This debate is not usually based on the ethical arguments used on Earth. The case for a plant-based diet on Mars is usually based on the fact that it will be impractical to raise large animals on Mars for the foreseeable future. However, how will the prospect of lab-grown meat on Mars impact this debate—particularly if you are able to consume meat without killing an animal? Regardless, while few of these products are being produced specifically for space, they could offer a revolutionary solution to many of the challenges facing future space explorers.

Since this process literally uses life to manufacture various products, it could also grow the needed biomass to be used for alcohol production. Rothschild believes that it may well be a less challenging task than trying to grow crops on Mars in more traditional ways. "Growing food on Mars is going to be somewhat of a challenge. There are things like perchlorates in the soil. But we do know bacteria on the Earth that deal with high levels of perchlorates. And they've got detoxifications mechanisms. Why not put those genes in the potato plant that the Martian use it...?" By doing so, "You could do it more efficiently off planet or maybe use biology for chemistry as we've been doing literally for millions of years, making alcohol, making material products."[76] Former NASA Ames Research Center Director Pete Worden echoes Rothschild's enthusiasm and believes that synthetic biology can present some impressive opportunities for future exploration and settlement. "With synthetic biology we may be able to produce groups more suitable to that environment and will be genetically modified for maximum efficiency and maximum flavor."[77]

Despite the Earth-based viability of this burgeoning technology, researchers have some concerns about how these processes will react to the various space environments, specifically the impact of variable levels of gravity. Will it work in Mars gravity, lunar gravity, and microgravity? Fortunately, we may have an answer to this question soon. As highlighted earlier in this chapter, the Eu:CROPIS (Euglena and Combined Regenerative Organic-Food Production in Space) satellite will simulate both lunar and Mars gravity. On board is NASA's PowerCell Payload, designed to examine the impact of these variations of gravity on mini-ecologies of microbes. This experiment will allow researchers to better understand how space environments could impact the processes required for synthetic biological research and production.

## What Drink Options Would Come with These Crops?

Assuming that human settlements with flourishing space farms have been built by the end of this century, what types of alcohol will be preferred and/or practical to manufacture? On Earth, a small number of crops have become the standard ingredients for alcoholic beverages, but can we expand our repertoire from which we produce alcoholic beverages? According to Penelepe Boston, "One of the things that I think is really interesting is that we have a gigantic number of plants on Earth that have not been traditionally used in food and spirit making. As we move forward and try to develop agricultural processes for a Martian base, I think it behooves us to look at that much greater palette of natural plant and organics which is quite remarkable. I think the potential is vast."[78] There will undoubtably be many more options for alcohol production, but what follows is a partial list of plants that are being considered as "space crops"—and the varieties of alcohol that can be produced from them.

### POTATOES

Originating in Peru, potatoes were brought to Europe by the Conquistadors and eventually spread to most parts of the world. In addition to providing a nutritious food option for future settlers residing in space colonies, potatoes are also known to be the staple crop for some popular varieties of alcoholic beverages. Perhaps the best known of these is potato vodka. While traditional vodkas in Russia and Poland were made from wheat, rye, and other grains, as potatoes became an inexpensive crop in eastern Europe, Ireland, and elsewhere, they became increasingly used to make vodka. Potatoes can also be used to make schnapps, whiskey, and other spirits.

## LETTUCE

Lettuce has been a successful crop on the International Space Station. It has proven to be a relatively easy crop to grow and can provide fresh sustenance for crew members on space missions. Lettuce does not usually come to mind when contemplating fermentable crops, but has some tradition in the world of alcoholic beverages. One such drink is Shochu. This Japanese beverage is usually made with such plants as rice, barley, sweet potatoes, buckwheat, or sugar cane, but at an "all-you-can-drink" Shochu bar in Tokyo, Japan, called HAVESPI, they have been making Shochu with unusual ingredients, including lettuce. According to the bar owner, they "opted to start with the lettuce Shochu to begin with, drinking it over ice as the staff recommended. Despite the high alcohol content, the taste was surprisingly smooth. Much like lettuce itself, the flavor was faint, refreshing and somehow cooling."[79]

## RADISHES

Radishes are yet another unlikely crop to ferment into alcoholic beverages, but a Japanese grocery store is marketing a wine make out of daikon—essentially an oversized radish. Called "Nerima Daikon," this sparkling wine was created by the Kubota Store using locally grown daikon. According to RocketNews24 Japan, "After getting a couple of bottles, our writer cracked one open and poured himself a nice big glass. It was a milky-white color with lots of tiny bubbles—and actually looked pretty good. But how would it taste? The daikon flavor is clearly there, but it wasn't overpowering, and the wine was very easy to drink. The flavor of the daikon went really well with the bubbles, and he found himself drinking the entire bottle without a thought."[80]

## SOY BEANS

Soy is fermented regularly on Earth for such products as soy sauce and other sauces and pastes used widely around Asia. There are also some references to soy beans being used in a soy wine, but this drink also uses other ingredients such as rice. However, in 2007 a soy-derived vodka was released called 3 Vodka, which advertising claimed was "the first time in history that soy has been distilled."

More recently, in 2017 the National University of Singapore (NUS) was able to turn a waste product of tofu, known as "whey," into an alcoholic beverage they call Sachi. Since this waste product has the potential of harming the environment, the team at NUS decided to see if they could find a beneficial use for it. According to PhD student Chua Jian-Yong, "I had previously worked on alcohol fermentation during my undergraduate studies in NUS, so I decided to take up the challenge of producing an alcoholic beverage using the whey. The drink turned out to be tasty, which is a pleasant surprise."[81]

## RICE

Rice has been known for millennia as a major crop for alcohol production. As mentioned earlier, rice was experimented with as part of a student experiment on Skylab in the 1970s, and the Chinese successfully grew rice on their Tiangong-2 space station. Perhaps the best-known rice-derived alcoholic beverage is sake, which is a Japanese rice wine. There are also spirits produced with rice including shoshu, which is produced with a sake-like mash and is sold through various brands around Asia and beyond. Rice is also used as the primary ingredient in making some beers, vodkas, and even whiskey.

## ALGAE

Algae is well known to be a productive feed stock for the production of ethanol and is also used for making biofuels and fuel supplements. In addition, algae are extremely nutritious, with many varieties having very high levels of protein and vitamins. Certain types of algae have also been made into consumable alcohol. For example, a drink producer in Wales has created a gin manufactured from seaweed, but seaweed is not likely to be a viable algae option in space.

However, if algae become a preferred crop in space for its various uses, it is probably just a matter of time before it is utilized to produce alcoholic beverages for future space settlers, since it is already used on Earth to produce ethanol. Even though it is not common to produce consumable alcohol from algae, it may present an interesting opportunity for fermentation in space.

## DATE PALMS

While they may be too large for short-term use, researchers in the United Arab Emirates (UAE) have investigated whether date palms would be suitable for a space-based plant. If this proves to be the case, it will also provide an ideal crop for the production of alcohol. Both the fruit and the sap from these trees are used to ferment alcohol. According to *Encyclopedia Britannica* (online), "Syrup, alcohol, vinegar, and a strong liquor are derived from the fruit." The sap is also often used to create palm wine. According to *The Drunk Botanist*, this wine can also be distilled into a spirit called arrack, "a general term referring to spirits made from sugary sap."

## CARROTS

Carrots are another crop that one typically does not think of as a feed stock for alcoholic drinks, but that too is changing. A Pennsylvania craft distillery called Board Room Spirits has created a 92 proof vodka entirely distilled from carrots. According to co-owner and co-founder Marat Mamedov, "We use literally a ton of carrots, ground down, fermented and distilled to capture the flavor and essence of that vegetable." Mamadov added, "Vodka is basically scentless and tasteless, whereas a distilled spirit made with a beet or a carrot really captures the essence and the experience of taking a vegetable or fruit, smelling it, biting into it and getting that flavor."[82]

## TOMATOES

Tomatoes have been a major staple of cuisine around the world for centuries, and it is therefore somewhat surprising that they have not become the basis of many alcoholic beverages. Tomatoes have been used in Japan for tomato sake and shochu, but now a tomato-based spirit has arrived. An example includes Black Tomato Spirit, a gin that is made from tomatoes and is produced by VOC Spirits. Laurent Cazottes also produces a liqueur made from 72 varieties of tomatoes appropriately called '72 Tomatoes. If tomatoes prove to be a successful crop in real-life space settlements, they may become the source of numerous new forms of drinks.

## Conclusion

Plant life and human expansion go hand in hand. This is the case whether the primary goal is to feed future settlers, to maintain their mental health, or to ferment space booze. It is essential for mission planners and the architects of settlements to understand this fact early on in the planning phases of their endeavors.

This fact is already obvious to the architects of a Mars-themed facility currently being planned in Las Vegas called Mars World that will simulate a

Mars city of the future. In the backstory of this themed facility, Mars World co-founder John Spencer explains that "there is a large farm underground to protect it from radiation and micrometeorites. And in the promenade in the tourist area it's a biosphere where there are fruit trees, plants, and farming areas on the walls. The Martians use just about every square inch for green areas, flowers, fruits and vegetables. It's all part of their culture that they're living within their farm."[83]

If the Mars World model turns out to be even close to accurate, Pete Worden thinks it is highly likely that some of the agricultural output will be devoted to alcoholic beverages. Worden stated, "Every culture on Earth [has] some sort of plant-based stimulant. Wherever they are they will want more than just the bare necessities. We're not just machines that need to be fueled."[84]

The question perhaps remains not "if" production is inevitable, but how long it will take before indigenous production becomes commonplace. Mike Dixon of the University of Guelph believes it will actually be a significant amount of time before "home-brews" will be possible. "I think the early occasions of alcohol on Mars exploration missions will be the stuff that people smuggle on board—just as the Russians have done in the past. We have a long way to go to see how plants will acclimate to the weird environment—how do they like the radiation environment? How do successive generations cope with any mutations? There are generations of questions to be answered in plant biology in space."[85] Whether Dixon is correct, or whether others who believe that settlers will ferment alcohol early on in agricultural pursuits are correct, few doubt that strong drink will be an inevitable byproduct of space farms.

# 7

# Final Thoughts

*"Ethanol is one of dozens of organic molecules found in space. Add them all up from one region of the galaxy to the next, and you will have a well-stocked bar. But don't stop there. Your cosmic [mix] will also contain the basic ingredients for life itself."*
—Neil deGrasse Tyson[1]

Alcohol is present naturally in space. In fact, there is an enormous cloud of alcohol 6,500 light years from Earth in a region of space that has the less-than-inspiring name, W3(OH). This intoxicating cloud surrounds a stellar nursery, a location that is literally the birthplace of new stars. Even if humanity could reach this interstellar distillery (at the speed the Space Shuttle traveled, it would take over 240 million years to get there), intergalactic drinkers might be disappointed that most of this cloud consists of methanol, rather than ethanol, and it would be toxic for human consumers. But W3(OH) is not the only source of naturally occurring alcohol in space. Another alcohol cloud resides in the constellation Aquilla and is "1,000 times the diameter of our entire Solar System, and contains enough alcohol to supply 300,000 pints (a pint = 473 ml) of beer every day to every single person on Earth for the next billion years."[2] Nevertheless, this interstellar liquor supply is over 10,000 light years from Earth, thus it would take over 350 million years to reach at Space Shuttle speed.

Clearly, humanity will not be tapping these massive supplies of alcohol that are scattered around the universe anytime soon. If we want to acquire sustainable supplies of alcohol for our own use in space, we will either need to manufacture it in space or ship it from Earth at great cost.

The stories highlighted in this book have shown that some remarkable projects have already taken place that may enable substantial manufacturing of liquor away from Earth. As impressive as these efforts have been, however, there is still some opposition to the prospect of ever legitimizing alcohol consumption in space. The reasons for these objections largely mirror concerns and criticisms of alcoholic beverages that have existed on Earth for centuries. These include:

## SAFETY

Obviously, it would be inadvisable to allow anyone to consume alcohol while performing tasks that could endanger the lives of others. Anyone who is responsible for piloting spacecraft or maintaining vital systems should not be allowed to drink—particularly within a reasonable time period before they perform their duties. However, this is no different than the rules that exist today on Earth. Pilots are not allowed to drink within eight hours of flying and astronauts are not allowed to drink within twelve hours of going into space. While it is true that space is inherently a far more dangerous place than the surface of the Earth, the question remains whether this prohibition should apply to tourists staying at space hotels, to potential settlers, or even government explorers who may stay on the surface of the Moon or Mars for years at a time. This is highly unrealistic, particularly with regard to privately funded activities.

## HEALTH

Health benefits versus the dangers of alcohol is a topic that has been debated for centuries. In recent years, numerous studies have claimed that moderate consumption of red wine and other alcoholic beverages is beneficial to the cardiovascular system and other elements of human health. However, a 2018 report called "Alcohol use and burden for 195 countries and territories, 1990–2016: a systematic analysis for the Global Burden of Disease Study 2016" that was published in *The Lancet*, a peer-reviewed medical journal, garnered an enormous amount of press. The authors of this report stated that no amount of alcohol is safe, which appeared to contradict some previous well-publicized studies. This report did not conduct a new study but reviewed and merged the results of hundreds of previous studies and data points. Based on a blind observation of data, it was shown that even small amounts of alcohol had a slight statistical negative impact on health. It should be noted, however, that these statistics did not consider other variables that may also have impacted health (for example, if they also smoked or took drugs).

While the data clearly indicated that heavy drinking increases chances for mortality (a fact that has not been disputed for centuries), the report only points to some very slight statistical increases potentially related to alcohol consumption. For example, if a sampling of 100,000 people who do *not* drink were observed, 914 of them would be expected to develop potentially fatal diseases like liver cancer, breast cancer, and numerous other serious ailments. However, if that same group of people consumed one drink a day, 918 would be expected to be diagnosed with one of these illnesses—an additional 4 out of 100,000. This number goes up to 977 at two drinks a day.

Assuming that this increase was in fact caused primarily by alcohol, it represents an extremely small danger. To put this into perspective, David Spiegelhalter, the Winton Professor for the Public Understanding of Risk at the University of Cambridge, noted that based on the data an individual would need to consume an additional 16 bottles of gin every year to add one more alcohol-related symptom. "That's a total of 400,000 bottles of gin among 25,000 people being associated with one extra health problem. Which indicates a rather low level of harm in these occasional drinkers." Spiegelhalter also noted that "there is no safe level of driving, but governments do not recommend that people avoid driving. Come to think of it, there is no safe level of living, but nobody would recommend abstention."[3] To expand this argument even further, compared with the inherent dangers of space exploration, any additional risk of responsible drinking would be infinitesimal.

In view of these dueling studies and reports, it is difficult to determine whether moderate or light alcohol consumption has a net positive or negative impact on human health. That said, the wisdom of many centuries still appears to be valid that moderate alcohol consumption poses few risks, but has social and psychological benefits that will be essential for space settlers.

## PUBLIC PERCEPTION

The perception that astronauts may be drunk in space—whether there is any truth in that at all—should not be dismissed. As discussed in controversy that arose as a result of the Bible reading during Apollo 8 or the decision to exclude wine from the Skylab menu, public opinion and perception can have a real impact. With regard to alcohol in space for longer-term missions, Jeffrey Kluger from *Time Magazine* stated, "I suspect the public would be of two minds. The majority would say 'rock on kids! You're in the service of our country and we trust you.'" However, he also believes a lot of people "will become huffy and self-righteous about it. We saw this in Apollo 14 when Al Shepard hit a golf ball on the Moon. This guy and the crew risked their lives to go to the Moon for their country and humanity. They nailed a perfect landing. They flew a flawless mission. And partly out of a physics question and partly out of fun, Al Shepard hit a ball. Flash forward forty-eight years and people are still saying 'We spent all that money to go play golf on the Moon.'"[4]

## PRACTICALITY

When humans begin exploring and settling space in earnest, virtually no one will dispute the fact that alcohol is far from a top priority. The pioneers must first learn to grow crops and to survive in harsh environments as well as accomplish their missions. However, anyone who believes (or expects) that space settlers will restrict themselves *only* to activities that are practical—or necessary for survival—has little understanding of human nature or what it is to be human. Former NASA astronaut Clayton Anderson even thinks this applies to long-term government-funded missions. "I definitely agree with the premise that if you're going to go on a six- to nine-month journey where you're probably going to end up living a year or two—before you can ever think of coming home—you kind of want to bring home with you. For many people, adult beverages are part of that lifestyle."[5] After all, if we only lived our lives conducting tasks that were practical, we would all live extremely dull lives. In short, life would not be worth living—and few of the great accomplishments of human history would ever have occurred, because those efforts probably would have been seen as impractical at the time.

Once again, one should not diminish the risks and dangers associated with heavy consumption of alcohol, but it would be far more irresponsible to deny the role that liquor has played in society and will likely continue to play in the future. A space-based teetotaling "utopia" is far from realistic, and when similar social experiments of this kind have been attempted on Earth, they have usually failed miserably. One of the best examples of a failed attempt to create an alcohol-free society occurred in Australia. During the 1780s, a gentleman named Thomas Townshend, Lord Sydney, became the British Secretary of State. Lord Sydney not only suggested a British colony in Australia, but also believed that in addition to hard work, "there would be no alcohol and no money—because without those there would be no crime."[6] Lord Sydney had originally planned for all inhabitants of this colony to be alcohol-free, including the guards, the marines, sailors, and others, but he had to back down on this demand as a result of extreme displeasure from these individuals. As for the elimination of crime, this too failed. According to Forsyth, during the early years of this colony, "the majority of all crimes in the colony would be either attempts to steal the precious drink, or violence committed under the influence."[7] Over time, because of these restrictions and corruption, rum became one of the most powerful unofficial currencies of Australia. In the end, the attempt to create a utopia based on Prohibition had exactly the opposite effect—and while Lord Sydney originally envisioned a society with no alcohol and no money, eventually alcohol *became* money in that remote colony.

To ignore these lessons—believing that humanity has grown beyond such "barbaric" needs—will merely invite a repeat of past failures. And while there are certainly legitimate and appropriate reasons for government space programs to officially restrict consumption of alcohol in space, if these restrictions are extended too far, they could end up having the opposite effect.

What are the challenges and opportunities over the next few decades for those interested in manufacturing and consuming alcohol in space?

## *Challenges*

• Government Resistance: The restriction on alcohol consumption in space will probably not extend to private facilities in space (except for the people responsible for the lives of others), but it remains an obstacle to experimentation in the space environment that involves consumption of alcohol— even to study how humans metabolize alcohol in space. This restriction is also not likely to be lifted any time soon. For safety and public relations concerns, government space facilities will likely remain dry for the foreseeable future.

• Resources: A sustainable colony or settlement—one that has a capacity for growth—requires reliable indigenous resources. The same requirements exist for any significant and sustained alcohol production. Otherwise, the ingredients will need to be shipped from Earth. If, for example, Budweiser is truly serious about manufacturing beer on Mars or in microgravity one day, this means that space farming will need to have matured enough to generate enough surplus to justify alcohol production.

• No Spacefaring Customers: No matter how much progress alcohol manufacturers and researchers make, none of their efforts matter without a market—without actual customers in space. Assuming that government space efforts will "officially" remain alcohol-free, the market will depend entirely on commercial/non-governmental activities. A company like Budweiser will certainly gain publicity and potentially advance some beneficial technologies, but the aspirational goal of becoming the first beer consumed or manufactured on Mars will remain an aspiration until there are a substantial number of people living and working in space. Companies like Budweiser, Suntory, and Ardbeg have little control over the speed and size of human settlement in space—their potential future market.

• Metabolizing Alcohol: There still have been no studies in space to determine how humans metabolize alcohol in space. We do know that consumption of liquor has taken place in space for decades, but since these activities are not officially acknowledged by space agencies, no studies have taken place to examine how cosmonauts and astronauts are impacted by alcohol in that environment. According to space medicine expert Saralyn Mark, "We also don't know how people are going to respond to alcohol in space. They may have a different response in space vs a terrestrial model. We don't know if they get combative. We don't know if they just go to sleep. We don't know if they get depressed. We just don't know how they're going to respond."[8] Nevertheless, based on anecdotal data, it is highly probable that alcohol does not have a dramatically different impact on humans in space, but this will not be scientifically proven or disproven until actual in-space studies are conducted.

## *Opportunities*

• Commercial Spaceflight Partnerships: Commercial spaceflight, including space tourism, appears to finally be on the verge of enabling access to space for non-government individuals. As these programs develop, alcohol producers and others with ambitions for testing their products, attempting space-based production methods, or researching the impact of alcohol on human physiology in space should build long-term partnerships with commercial players to enable activities that are not possible or cost effective through government space facilities.

• Cross-over collaboration/benefits to Earth: Alcohol producers have little ability to impact the development of launch vehicles and crew habitats, but they can contribute real value to essential technologies and capabilities that are required for sustainable human settlement in space, as well as for alcohol production. First and foremost, among these overlapping needs is agriculture. Partnering or investing in new farming techniques or methods of growing crops in extreme environments will benefit the future of space exploration, enable alcohol manufacturing, and potentially provide real benefits here on Earth. These collaborations could also include methods for utilizing indigenous resources, including water. There is also the potential for innovation that goes well beyond satisfying the need for a drink. As Bill Nye noted, "And who knows what we'll learn about beer making by trying to do it in low gravity. Who knows what innovation may emerge?"[9]

• Professional Association: With an ever-growing international community of professionals and citizen scientists advancing the concept of space alcohol, the time has probably come for a non-profit professional association. Such a group would not only bring these interested parties together, but also serve as a hub to assemble manufacturers, researchers, citizen scientists, and experts in other relevant fields of knowledge. This association could be based on a model similar to the Home Brewers Association and other comparable organizations.

We live in a remarkable age. For centuries, science fiction authors, futurists, and others have hypothesized and prognosticated about building space settlements, sending humans to Mars, and expanding human civilization beyond Earth. These dreams appear to finally be on the verge of becoming reality. As space becomes accessible to ordinary citizens for pleasure and settlement, there will be a growing demand to provide space tourists with some of the pleasures they would expect on Earth. This includes alcoholic beverages. As Bill Nye noted, "If there's a fancy restaurant with an enormous window and a view of the Earth below, you'd want to serve some fine wine, and I can imagine a flight of wines or a flight of beers, little shots of it, so as you flew over each continent you would have the beverage from that place, you know, your 90 minute orbit. I can easily imagine that there'd be restaurants in space of the future."[10]

This is quite consistent with the predictions of Mark Forsyth, who believes that "we shall, as a species, down our final earthly noggins, stumble into our spacecraft and leave behind this little ball of rock. It will be a great journey. As we break above the atmosphere leaving this old earth behind us, the gods will be there to cheer us on…. And we shall zoom, drunkenly, into the infinite."[11]

Perhaps we should not "zoom drunkenly into the infinite," but we also should not zoom forward into denial or believe that sending humans to a new location will change human nature. As we saw in the case of the British settlement of Australia, it would almost certainly be doomed to fail. Over the next century, as humanity expands further into our solar system, new customs and social experiments will undoubtedly materialize. If all goes as planned, this expansion will be based on self-determination and freedom. As is the case on Earth, there unquestionably will be settlements or colonies that decide to remain "dry," but it is almost certain that there will be many others who will look off to a pale blue dot in space and raise their glasses and say "Cheers!"

# Chapter Notes

## Introduction

1. Neil deGrasse Tyson (astrophysicist), in an email discussion with the author, July 2017.
2. Andy Weir (author of *The Martian* and *Artemis*), in an email discussion with the author, June 2017.
3. Simon "Pete" Worden (former director, NASA Ames Research Center), in a discussion with the author, January 2018.
4. Iain Gately, *Drink: A Cultural History of Alcohol* (New York: Gotham Books, 2008), 1.

## Chapter 1

1. Michael Richards, "Alcohol Abuser," International Churchill Society, https://Winston Churchill.org.
2. Andrew Curry, "Our 9000-Year Love Affair with Booze," *National Geographic* magazine, last modified February 2017, https://www.nationalgeographic.com/.
3. Reed Walker (co-owner, Cotton & Reed Distillery), in an email conversation with the author, March 2018.
4. Ian O'Neill (astrophysicist; science communicator), in an email conversation with the author, March 2019.
5. Patrick E. McGovern, et al., "Fermented Beverages of Pre- and Proto-Historic China," *Proceedings of the National Academy of Sciences*, last modified December 2004, https://www.pnas.org/.
6. Robin Dunbar, "Why Drink is the Secret to Humanity's Success," *Financial Times*, last modified August 10, 2018, https://www.ft.com/.
7. McGovern, "Fermented Beverages."
8. Ian S. Hornsey, *Alcohol and Evolution of Human Society* (Cambridge, U.K.: RSC Publishing, 2012), 1.
9. Mark Forsyth, *A Short History of Drunkenness* (New York: Three Rivers Press, 2017), 26.

10. Iain Gately, *Drink: A Cultural History or Alcohol* (New York: Gotham Books, 2008), 5.
11. Gately, *Drink*, 5.
12. Gately, *Drink*, 7.
13. "King Tut Liked Red Wine," *American Chemical Society* press release, last modified March 15, 2004, https://www.eurekalert.org/.
14. Gately, *Drink*, 11.
15. Forsyth, *A Short History*, 52.
16. Forsyth, *A Short History*, 57.
17. Hugh Johnson, *The Story of Wine* (London: Mitchell Beazley, 1989), 27.
18. Johnson, *The Story of Wine*, 27.
19. Johnson, *The Story of Wine*, 27.
20. Gately, *Drink*, 31.
21. Gately, *Drink*, 31.
22. Johnson, *The Story of Wine*, 36.
23. Gately, *Drink*, 45.
24. Joe Clark, Stuart Derrick, *The Ultimate Book of Whiskey* (New York: Paragon Books Ltd., 2014), 8.
25. Gately, *Drink*, 41.
26. Marni Davis, *Jews and Booze: Becoming American in the Age of Prohibition* (New York: NYU Press, 2012), 6.
27. Khaled Diab, "A Secular Muslim's Guide to Drinking During Ramadan," *The Washington Post*, last modified May 31, 2018, https://www.washingtonpost.com/.
28. Edward William Lane, *An Account of the Manners and Customs of Modern Egyptians* (London: John Murray, Albemarle Street, 1860), 94.
29. Alice Gordenker, "Sake Barrels in Shrines," *The Japanese Times*, last modified October 16, 2007, https://www.japantimes.co.jp/.
30. Daimon Yasutaka, "Sake—Drink of the Gods, Drink for the People" lecture at Japan Society, New York, NY, October 12, 2000, http://esake.com/Brewers/DaimonB/Naorai/naorai.html.
31. Yasutaka, "Sake—Drink of the Gods."
32. Julia Blakely, "Beer on Board in the Age of Sail," *Smithsonian Libraries Unbound*, last

modified August 2, 2017, https://blog.library.si.edu/.

33. Gately, *Drink*, 95.

34. Stanton Peele and Archie Brodsky, "Exploring Psychological Benefits Associated with Moderate Alcohol Use," *Drug and Alcohol Dependence* 60 (2000): 221–247.

35. John Livingstone-Learmonth, *The Wines of the Rhone* (3rd Edition) (London: Faber and Faber, 1992), 89.

36. Noelle Plack, "Intoxication and the French Revolution," *Age of Revolution*, last modified December 5, 2016, https://ageofrevolutions.com/.

37. Mikhail Butov, "When the Tsar Banned Booze," *Russia Today*, last modified August 18, 2014, https://www.rbth.com/.

38. Gately, *Drink*, 122

39. Gately, *Drink*, 123.

40. Increase Mather, *Wo to Drunkards* (Boston: Tomothy Green, at the Lower End Middle-Street, 1712), 7.

41. Mather, *Wo to Drunkards*, 7.

42. Bruce Bustard, *Spirited Republic: Alcohol's Evolving Role in U.S. History*, last modified 2014, https://www.archives.gov/.

43. Benjamin Rush, *An Inquiry into the Effects of Ardent Spirits upon the Human Body and Mind* (New-York: Printed for Cornelius Davis, 1811).

44. Rush, *Ardent Spirits*, 1.

45. Rush, *Ardent Spirits*, 13.

46. Rush, *Ardent Spirits*, 2.

47. Rush, *Ardent Spirits*, 2.

48. Rush, *Ardent Spirits*, 2.

49. "Thomas Jefferson to John F. Oliveira Fernandes (December 16, 1815)," *The Papers of Thomas Jefferson. Retirement Series. Volume Nine*, Encyclopedia Virginia, https://www.encyclopediavirginia.org/Jefferson_to_Oliviera_1815.

50. "Extract from Thomas Jefferson to Jean Guillaume Hyde de Neuville (December 13, 1818)," Jefferson Quotes & Letters, The Jefferson Monticello, http://tjrs.monticello.org/letter/364.

51. "Philip Mazzei to George Washington (January 27, 1779)," *Washington Papers*, https://founders.archives.gov/documents/.

52. "Thomas Jefferson to William Alston (October 6, 1818)," *The Papers of Thomas Jefferson: Retirement Series* (Princeton: Princeton University Press, 2016), 302.

53. Walter Isaacson (edited and annotated), *A Benjamin Franklin Reader* (New York: Simon & Schuster Paperbacks, 2003), 295.

54. "Benjamin Rush's Rules on Health," from the *Letters of the Lewis and Clark Expedition with Related Documents*, 1978, https://www.lewis-clark.org/article/2394.

55. "Benjamin Rush's Rules."

56. "September 14, 1803," *Journals of the Lewis & Clark Expedition*, https://lewisandclarkjournals.unl.edu/item/lc.jrn.1803-09-14.

57. Robert R. Hunt, "Lewis's Early Army Experience," 1991, last modified http://www.lewisclark.org/article/2505.

58. Henry Jeffreys, "Shackleton's Booze Kept the Men in High Spirits," *The Guardian*, last modified September 11, 2015, https://www.theguardian.com/.

59. Roland Huntford, *Shackleton* (New York: Carroll & Graf, 1998).

60. Kari Herbert, *Polar Wives* (Vancouver: Greystone Books, 2012), 322.

61. Charles McGrath, "Spirits of the South Pole," *The New York Times Magazine*, last modified July 21, 2011, https://www.nytimes.com/.

62. Rush, *Ardent Spirits*, 24.

63. "Roots of Prohibition," *Prohibition: A Film by Ken Burns and Lynn Novick*, last modified 2011, http://www.pbs.org/kenburns/.

64. Davis, *Jews and Booze*, 2.

65. Nick Hines, "How the Church Saved the Wine Industry During Prohibition," *VinePair*, last modified December 20, 2016, https://vinepair.com/.

66. "The History of Alcohol Prohibition," Shaffer Library of Drug Policy, http://www.druglibrary.org/.

67. Cordelia Hebblethwaite, "Absinthe in France: Legalising the 'Green Fairy,'" last modified May 4, 2011, https://www.bbc.com/news/.

68. Jason Harrell, *The Bartending Therapist* (Amazon Digital Services, 2016), 3.

69. Jon Sharman, "Going to the Pub is Officially Good For You, According to Oxford University Researchers," *Independent*, last modified January 7, 2017, https://www.independent.co.uk/news/.

70. Susan M. Barbieri, "Shrinks Who Work for Tips," *Chicago Tribune*, last modified December 8, 1991, https://www.chicagotribune.com/.

71. Harrell, *The Bartending Therapist*, 6.

72. Noreen Malone, "Undercover at the U.N. Lounge, Where Diplomats Get Drunk and Handsy," *The New Republic*, last modified January 30, 2014, https://newrepublic.com/.

73. Malone, "Undercover at the U.N."

74. Malone, "Undercover at the U.N."

75. Ian O'Neill interview, March 2019.

76. Joey Bunch, "Rep. Soggy Sweat could guide Colorado lawmakers on whiskey," *The Spot, The Denver Post*, last modified April 26, 2016, http://blogs.denverpost.com/thespot/.

77. Dunbar, "Why Drink is the Secret."

## Chapter 2

1. H.G. Wells, *The First Men in the Moon* (Indianapolis: The Bowen-Merrill Company, 1901), 111.

2. Wells, *The First Men*, 111.

3. Wells, *The First Men*, 112.

4. Jules Verne, *From Earth to the Moon* (New York: Charles Scribner's Sons, 1890), 204–205.

5. Ray Bradbury, *The Martian Chronicles* (New York: Simon & Schuster, 1977), 209.

6. Bradbury, *The Martian Chronicles*, 216.

7. Gregory Benford, "All the Beer on Mars," *Isaac Asimov's Scientific Fiction* vol. 13, no. 1 (January 1989): 61.

8. Benford, "All the Beer on Mars," 66.

9. Benford, "All the Beer on Mars," 75.

10. Gregory Benford (science fiction novelist), in a discussion with the author, December 2018.

11. David Brin (science fiction author), in an email discussion with the author, February 2018.

12. David Brin, "Talk Farm Dynamo," 1983, http://www.davidbrin.com/tankfarm.htm.

13. Gene Roddenberry, "Star Trek Judgement Rites—Gene Roddenberry Interview," YouTube video, 17:37, posted by Interplay Productions, December 2014, https://www.youtube.com/.

14. Jonathan Franks (William Riker, *Star Trek: The Next Generation*), in an interview with the author, June 2018.

15. Ryan McElveen (COO and sommelier, Silver Screen Bottling), in an interview with the author, December 2018.

16. McElveen interview.

17. McElveen interview.

18. John Spencer (Executive VP and Chief Designer, Mars World Enterprises), in an interview with the author, March 2018.

19. J.C. Reifenberg (co-owner, Scum & Villainy Cantina), in an interview with the author, December 2018.

20. Reifenberg interview.

21. Reifenberg interview.

22. Frakes interview.

23. Robert Picardo ("The Doctor," *Star Trek Voyager*), in an interview with the author, June 2018.

24. "J. Michael Straczynski's Origin Story," *San Diego City Beat*, last modified, September 5, 2012, sdcitybeat.com.

25. J. Michael Straczynski, Twitter post, July 11, 2018, 12:49 AM, https://twitter.com.

26. Spencer interview.

27. Brin interview.

28. Spider Robinson, *The Callahan Chronicles* (New York: A Tom Doherty Associates Book, 1997), 388.

29. Gregory Benford, "A Fairhope Alien," *Stories of the Blue Moon Café II*, edited by Sonny Brewer (San Francisco: MacAdam/Cage, 2003), 134.

30. Benford interview.

31. Douglas Adams, *The Hitchhiker's Guide to the Galaxy* (New York: Ballantine Books, 2005—Kindle edition), 7.

32. Adams, *The Hitchhiker's Guide*, 2.

33. Adams, *The Hitchhiker's Guide*, 12.

34. Adams, *The Hitchhiker's Guide*, 12.

35. Andy Weir, "A conversation with #1 New York Times bestselling author of The Martian, Andy Weir, about his exciting new novel," http://www.andyweirauthor.com/books/artemis-tr.

36. Andy Weir, *Artemis* (New York: Crown, 2017), 32.

37. Andy Weir (author, *The Martian* and *Artemis*), in an email discussion with the author, June 2017.

38. Weir, *Artemis*, 35.

39. Weir, *Artemis*, 35.

40. Weir interview.

41. O'Neill interview.

42. Douglas Adams, *The Restaurant at the End of the Universe* (New York: Ballantine Books, 2005), 113.

43. Adams, *The Hitchhiker's Guide*, 12.

44. Adams, *The Hitchhiker's Guide*, 9.

45. Benford interview.

## Chapter 3

1. John Grunsfeld (astronaut, Space Shuttle; former chief scientist and associate administrator, NASA), in an interview with the author, August 2018.

2. Clayton Anderson (astronaut, Space Shuttle), in an interview with the author, June 2018.

3. Anderson interview.

4. Bryan Lufkin, "Why Astronauts are Banned from Getting Drunk in Space," *BBC*, last modified February 20, 2017, http://www.bbc.com/.

5. "Just Like Home," last modified April 10, 2009, https://www.nasa.gov/audience/.

6. Colin Burgess, *Sigma 7: The Six Mercury Orbits of Walter M. Schirra, Jr.* (Chichester, UK: Praxis Publishing, 2016), 185.

7. Burgess, *Sigma 7*, 185.

8. Frank Borman, interview with Catherine Harwood, April 13, 1999, in Las Cruces, New Mexico, Transcript, NASA Johnson Spaceflight

Center Oral History Project, https://www.jsc.nasa.gov/history/oral_histories/.

9. "Apollo 8 Flown Unopened Bottle of Coronet Brandy from the Personal Collection of Mission Command Module Pilot James Lovell….," last modified October 7, 2008, https://historical.ha.com/.

10. Buzz Aldrin with Ken Abraham, *Magnificent Desolation: The Long Journey Home from the Moon* (New York: Harmony Books, 2009), 26.

11. Yasmine Hafiz, "The Moon Communion Of Buzz Aldrin That NASA Didn't Want To Broadcast," *Huffington Post*, last modified December 6, 2017, https://www.huffingtonpost.com/.

12. Heidi Blake, "Apollo 11 Moon landing: top quotes from the mission that put man on the Moon," *The Telegraph*, last modified July 20, 2009, https://www.telegraph.co.uk/.

13. Hafiz, "The Moon Communion."

14. Hafiz, "The Moon Communion."

15. Yevgeny Levkovich, "Space smugglers: How Russian cosmonauts sneak booze into outer space," *Russian Beyond*, last modified April 12, 2017, https://www.rbth.com/.

16. Levkovich, "Space Smugglers."

17. Levkovich, "Space Smugglers."

18. Levkovich, "Space Smugglers."

19. Levkovich, "Space Smugglers."

20. Richard Garriott (private astronaut, visited the ISS in 2008), in an interview with the author, December 2018.

21. "The ISS Menu: Mayo, Espressos, Booze? Cosmonauts Reveal Their Secrets," last modified February 16, 2015, https://sputniknews.com/.

22. Levkovich, "Space Smugglers."

23. Saralyn Mark (physician; space medicine expert), in a conversation with the author, January 2019.

24. Grunsfeld interview.

25. Levkovich, "Space Smugglers."

26. Nicola Twilley, "Why Astronauts Were Banned From Drinking Wine In Outer Space," *GIZMODO*, last modified January 31, 2014, https://gizmodo.com/.

27. Charles T. Bourland and Gregory L. Vogt, *The Astronaut's Cookbook: Tales, Recipes, and More* (New York: Springer, 2010), 170–171.

28. Twilley, "Why Astronauts Were Banned."

29. Arthur Hill, "Astronauts Finally Get Out of this World Wine," *The Milwaukee Journal*, August 1, 1972.

30. Hill, "Astronauts Finally Get Out."

31. "PART III: Skylab Development and Operations: February 1970–November 1974," *SP-4011 Skylab: A Chronology*, August 10, 1972, https://history.nasa.gov/.

32. Twilley, "Why Astronauts Were Banned."

33. Bourland, *The Astronaut's Cookbook*, 171.

34. Charles Bourland (former NASA food scientist), in an email interview with the author, June 2017.

35. David Hilt, Owen Garriott, Joe Kerwin, *Homesteading Space: The Skylab Story* (Lincoln: University of Nebraska Press, 2008), 110.

36. Alan Boyle, "Alcohol in space? Da!," *NBC News*, last modified October 14, 2010, https://www.nbcnews.com/.

37. Jeffrey Manber (CEO, Nanoracks), in an interview with the author, February 2018.

38. Manber interview.

39. Manber interview.

40. Manber interview.

41. Manber interview.

42. Yeganeh Torbati, "U.S. and French Navies Differ on Alcohol Aboard Ships," originally by *Reuters*, last modified December 19, 2015, https://gcaptain.com/.

43. Torbati, "U.S. and French Navies."

44. Torbati, "U.S. and French Navies."

45. Grunsfeld interview.

46. Grunsfeld interview.

47. Grunsfeld interview.

48. Michael Foale, "What it's like to … Spend a Year Floating in Space," *The Telegraph*, last modified October 23, 2013, https://www.telegraph.co.uk/.

49. Foale, "What it's like to…."

50. Norman Thagard, "Interview with Rebecca Wright, Paul Rollins, Carol Butler, (September 16, 1998)," Shuttle-Mir Oral History Program, https://spaceflight.nasa.gov/history/.

51. Thagard oral history.

52. Thagard oral history.

53. Thagard oral history.

54. Thagard oral history.

55. Jane Anson, "Anson on Thursday: The Surprising Adventure of Lynch Bages in Space," *Decanter*, last modified June 18, 2015, https://www.decanter.com/wine-news/.

56. David Remnick, "For France's Astronauts, It's Pass the Pâté" *The Washington Post*, June 19, 1985, https://www.washingtonpost.com/archive/.

57. Anson, "Anson on Thursday."

58. Thomas Firh, "The Weightless World with Patrick Baudry," *les others*, last modified September 2, 2016, https://www.lesothers.com/.

59. Grunsfeld interview.

60. Marcia Dunn, "Fire, Collision, Noxious Fumes, and Cognac—All in a Day's Work on Mir," *L.A. Times*, last modified February 22, 1998, http://articles.latimes.com/.

61. Dunn, "Fire, Collision, Noxious Fumes."

62. Anderson interview.

63. Anderson interview.
64. Garriott interview.
65. Garriott interview.
66. Anderson interview.
67. Manber interview.
68. Anderson interview.
69. Mark Gerchick, "A Brief History of the Mile High Club," *The Atlantic* (January/February 2014), https://www.theatlantic.com/.
70. Grunsfeld interview.
71. Cheryl L. Mansfield, "If Walls Could Talk," last modified January 20, 2011, https://www.nasa.gov/.
72. Scott Kelly (astronaut, Space Shuttle), in a conversation with the author, December 2018.
73. Grunsfeld interview.
74. Scott Kelly, *Endurance* (New York: Knopf, 2017), 14–15.
75. Paul Rincon, "The First Spacewalk," *BBC*, last modified October 13, 2014, http://www.bbc.co.uk/.
76. Kelly, *Endurance*, 15.
77. Kelly, *Endurance*, 30–31.
78. Garriott interview.
79. "NASA Astronaut Health Care System Review Committee, February—June, 2007, Report to the Administrator," https://www.nasa.gov/.
80. Bryan O'Connor (Chief Safety and Mission Assurance Officer, NASA), *C-Span*, August 29, 2007, https://www.c-span.org/.
81. Lindsay Goldwert, "Russia Denies Drunk Astronaut Report," *CBS News*, last modified July 29, 2007, https://www.cbsnews.com/.
82. Simon Saradzhyan, "Russians Let Cosmonauts Drink—But They Mustn't Go Into Orbit," *The Telegraph*, Mast modified January 15, 2006, https://www.telegraph.co.uk/.
83. Olga Khazan, "On Getting Drunk in Antarctica," *The Atlantic*, last modified July 15, 2013, https://www.theatlantic.com/.
84. Khazan, "On Getting Drunk in Antarctica."
85. Tia Ghose, "Cold Comfort: Why People in Antarctica Are Such Boozehounds," *Live-Science*, last modified October 13, 2015, https://www.livescience.com/52467-why-antarctica-fuels-excess-drinking.html.
86. "HMP Alcohol Policy," Haughton Mars Project document, last modified July 1, 2018.

## Chapter 4

1. David Brin (futurist; science fiction author), in an email conversation with the author, February 2018.
2. Jeffrey Kluger (writer, *Time Magazine*), in a conversation with the author, May 2018.
3. Teagan Boda, "The Cola Space Race," *The Cola Wars: Creation of the Soda Giants*, last modified November 27, 2016, http://scalar.usc.edu/works/colawars/the-cola-space-race.
4. Richard Garriott (private astronaut, visited the ISS in 2008), in an interview with the author, December 2018.
5. Garriott interview.
6. Penelope Boston, David Blackmore, Reed Walker, Chris Carberry, "Libations in Space," Panel discussion, Explore Mars, Inc., Washington, D.C., May 11, 2017.
7. Marina Koren, "Everything You Never Thought to Ask About Astronaut Food," *The Atlantic*, last modified December 15, 2017, https://www.theatlantic.com/.
8. Clayton Anderson (astronaut, Space Shuttle), in a conversation with the author, June 2018.
9. Jim Romanoff, "When It Comes to Living in Space, It's a Matter of Taste," *Scientific American*, last modified March 10, 2009, https://www.scientificamerican.com/.
10. Garriott interview.
11. Romanoff, "When it Comes to Living."
12. Charles Bourland (former food scientist, NASA), in an email interview with the author, June 2017.
13. Bourland interview.
14. Kirsten Sterrett (former student, University of Colorado), in a conversation with the author, August 2018.
15. Sterrett interview.
16. Kirsten Sterrett, an interview with Sue Clark. "You Can Get It Orbiting…," *ABC* (Australia), last modified December 9, 2002, https://www.abc.net.au/.
17. Sterrett interview.
18. Kirsten Sterrett, Marvin Luttges, Steven J. Simske, et al., "A Study of Saccharomyces Uvarum Fermentation in a Microgravity Environment, *Technical Quarterly* vol. 3, no. 1 (1996): 33–38.
19. Sterrett, "A Study of Saccharomyces," 33–38.
20. Sterrett, "A Study of Saccharomyces," 33–38.
21. "You Can Get it Orbiting."
22. Sterrett, "A Study of Saccharomyces," 33–38.
23. Sterrett interview.
24. "Suds in Space," last modified September 21, 2001, https://science.nasa.gov/.
25. "Suds in Space."
26. "Suds in Space."
27. Sterrett interview.

28. Sterrett interview.
29. "The Making of the Legendary Biru Re-size," https://sapporobeer.com/our-history/.
30. Manabu Sagimoto, Elena Shagimar-donova, Oleg Gusev, et al., "Gene Expression of Barley Grown in Space," *Space Utiliz Res* vol. 24 (2008), https://repository.exst.jaxa.jp/
31. "Sapporo Space Barley," Sapporo press release, December 3, 2009, https://sapporobeer.com/press/sapporo-space-barley/.
32. Danielle Demetriou, "Japan unveils 'space beer,'" *The Telegraph*, last modified December 3, 2008, https://www.telegraph.co.uk/.
33. Ricardo Marques (VP of Marketing, Bud-weiser), in an email conversation with the author, July 2018.
34. "This Bud's For Mars." Budweiser press release, March 13, 2017, https://www.prnewswire.com/news-releases/.
35. Gary Hanning (Director of Global Barley Research, Anheuser-Busch), in an email con-versation with the author, July 2018.
36. Budweiser press release.
37. Hanning interview.
38. Rosalie Chan, "Could Budweiser Actually Brew Beer on Mars?," *Inverse*, last modified March 13, 2017, https://www.inverse.com/.
39. Hanning interview.
40. Hanning interview.
41. Hanning interview.
42. Hanning interview.
43. "More than 20 U.S. National Laboratory Payloads Part of SpaceX's 16th Mission to Space Station," ISS U.S. National Laboratory press re-lease, November 27, 2018, https://www.iss-casis.org/.
44. Hanning interview.
45. Ian O'Neill interview.
46. Anderson interview.
47. Hanning interview.
48. Budweiser press release.
49. Jason Held (founder, Saber Astronautics), in a conversation with the author, April 2018.
50. Held interview.
51. Held interview.
52. "Tasting Notes," Nitro Stout, 4 Pines Brewing Company, https://4pinesbeer.com.au/our-beer/stout/.
53. Held interview.
54. Held interview.
55. Jason Held interviewed by Dr. Space Junk, "Vostok Space Beer: from Prehistory to Space Tourism. An Interview with Dr. Jason Held," last modified October 18, 2010, http://zoharesque.blogspot.com/2010/.
56. "How it Works," ZERO G, http://www.gozerog.com/index.cfm?fuseaction=Experience.How_it_Works.
57. Held interview.
58. "Astronauts4Hire Launches, Selects Initial Candidates for Commercial Scientist As-tronauts," Astronauts4Hire press release, April 12, 2010, http://www.astronautforhire.com/2010/04/astronauts4hire.html.
59. Brian Shiro (former President and CEO of Astronauts4Hire), in a conversation with the author, March 2018.
60. Held interview.
61. Held interview.
62. Held interview.
63. Held interview.
64. Held interview.
65. "Vostok Space Beer," Indiegogo campaign, 2018, https://www.indiegogo.com/projects/vostok-space-beer#/.
66. Dr. Space Junk interview.
67. Dr. Space Junk interview.
68. Jamie Floyd (co-founder, Ninkasi Brew-ing), in a conversation with the author, July 2018.
69. Floyd interview.
70. Floyd interview.
71. Elizabeth Howell, "Space Beer! New Brew Made from Spacefaring Yeast," last modified April 28, 2015, https://www.space.com/.
72. Floyd interview.
73. "Ninkasi Brewing Company Introduces Ground Control, Imperial Stout—Fermented with Space-traveled Yeast," Ninkasi press release, April 13, 2015, https://nsp.ninkasibrewing.com/.
74. Floyd interview.
75. Floyd interview.
76. Floyd interview.
77. Deborah L. Jude, "Can You Brew Beer on the Moon?" University of California, San Diego website, last modified January 19, 2017, https://www.universityofcalifornia.edu/news/.
78. Neeki Ashari (student, UC San Diego), in an email conversation with the author, July 2017.
79. Jude, "Can You Brew Beer on the Moon?"
80. Ashari interview.
81. Ashari interview.
82. Ashari interview.
83. Paul Bakken (principal investigator, MDRS Crew #149), in a conversation with the author, May 2018.
84. Kellie Gerardi (crew member, MDRS Crew #149), in an interview with the author, June 2018.
85. Gerardi interview.
86. Bakken interview.

87. Bakken interview.
88. Bakken interview.
89. Gerardi interview.

## Chapter 5

1. Jeffrey Manber (CEO, Nanoracks), in an interview with the author, February 2018.
2. Bill Lumsden, "The Impact of Microgravity on the Release of Oak Extractives into Spirit," Ardbeg white paper, 2015, https://www.ardbeg.com/sites/ardbeg.com/.
3. David Blackmore (brand ambassador, Ardbeg) in an email conversation with the author, March 2018.
4. Lumsden, "The Impact of Microgravity."
5. "Ardbeg Reveals Results of 'Space Whisky' Experiment," *BBC News*, last modified September 15, 2015, https://www.bbc.com/news/.
6. Manber interview.
7. Lumsden, "The Impact of Microgravity."
8. Penelope Boston, David Blackmore, Reed Walker, Chris Carberry, "Libations in Space," Panel discussion, Explore Mars, Inc., Washington, D.C., May 11, 2017.
9. "Libations in Space."
10. Blackmore interview.
11. Manber interview.
12. "Ardbeg Distillery Launches U.S. Rocket Tour Celebrating "World First" Space Experiment," Ardbeg press release, May 1, 2012, https://www.prnewswire.com/news-releases/.
13. Manber interview.
14. *BBC News* article, September 15, 2015.
15. Manber interview.
16. "Yamazaki Distillery: Yamazaki: Where it all Began in 1923," Suntory, https://house.suntory.com/en/na/distilleries/yamazaki.
17. "Elucidating the Mechanism Mellowing Alcoholic Beverage-Space Experiments to Begin Soon," July 30, 2015, https://www.suntory.com/news/2015/12432.html.
18. Suntory press release.
19. Kelly Dickerson, "A World-Famous Distillery is Launching Whiskey into Space for a Surprising Reason," *Business Insider*, last modified August 3, 2015, https://www.businessinsider.com/.
20. Kevin Conlon, "Booze Arrives at Space Station for Out-of-this-World Experiment," *CNN*, last modified August 25, 2015, https://www.cnn.com/.
21. Mumm Cordon Stellar website, https://www.mumm.com/en-au/mumm-grand-cordon-stellar.
22. Octave de Gaulle (space head designer), in a conversation with the author, July 2018.

23. de Gaulle interview.
24. de Gaulle interview.
25. Laura Seal, "Mumm to Launch Space Champagne for Astronauts," *Decanter*, last modified June 8, 2018, https://www.decanter.com/.
26. Seal, "Mumm to Launch Space Champagne."
27. Seal, "Mumm to Launch Space Champagne."
28. "With Mumm Grand Cordon Stellar, Life in Space Will Never Be the Same Again," Mumm press release, June 7, 2018, https://www.prnewswire.com/news-releases/.
29. Mumm press release.
30. de Gaulle interview.
31. de Gaulle interview.
32. Seal, "Mumm to Launch Space Champagne."
33. de Gaulle interview.
34. "Mumm Champagne presents: Mumm Grand Cordon Stellar Project," YouTube video, 1:48, Posted June 7, 2018, https://www.youtube.com/watch?v=VKxeaUkmFr4.
35. Mumm video.
36. Stephanie Mlot, "Toast to Space Travel With Zero-Gravity Champagne," Geekwww, last modified September 12, 2018, https://www.geek.com/.
37. Amie Ferris-Rotman, "White Wine on the Red Planet? Scientists in Georgia are Hunting for a Perfect Martian Grape," *The Washington Post*, last modified January 7, 2019, https://www.washingtonpost.com/.
38. Tom Phillips, "The Red Planet: China Sends Vines Into Space in Quest for Perfect Wine," *The Guardian*, last modified September 21, 2016, https://www.theguardian.com/.
39. Bill Nye "The Science Guy", in a conversation with the author, May 2018.
40. Ferris-Rotman, "White Wine on the Red Planet?"
41. Ferris-Rotman, "White Wine on the Red Planet?"
42. Ferris-Rotman, "White Wine on the Red Planet?"
43. Greg Olsen (private astronaut; owner, Olsen Wine), in a conversation with the author, July 2018.
44. Olsen interview.
45. Olsen interview.
46. Olsen interview.
47. Olsen interview.
48. Kluger interview.
49. Nye Interview.
50. Richard Garriott (private astronaut, visited the ISS in 2008), in an interview with the author, December 2018.

51. Samuel Coniglio (founder, Cosmic Lifestyle Corp), in an email discussion with the author, June 2018.

52. Coniglio interview.

53. Coniglio interview.

54. "Zero Gravity Cocktail Project," Kickstarter campaign by Cosmic Lifestyle Corp, 2015, https://www.kickstarter.com/.

55. "How Do You Make a Whisky Glass for Space?," last modified September 2, 2015, https://medium.com/space-glass/.

56. James Parr (founder, Open Space Agency), in a conversation with the author, October 2018.

57. Medium.com, "How Do You Make a Whisky Glass?"

58. Medium.com, "How Do You Make a Whisky Glass?"

59. Parr interview.

60. Parr interview.

61. Parr interview.

62. Medium.com, "How Do You Make a Whisky Glass?"

63. Zoe Cormier, "The End of Hangovers," last modified December 21, 2017, https://medium.com/neodotlife/.

64. "Alcosynth: The Hangover-Free Future of Drinking," YouTube video, 1:34, December 13, 2017, https://www.youtube.com/watch?v=J0-JEh5mZ7E.

65. Katie Forster, "'Hangover-Free Alcohol' Could Replace All Regular Alcohol by 2050, says David Nutt," Independent, last modified September 22, 2016, https://www.independent.co.uk/.

66. Cormier, "The End of Hangovers."

67. "Natural Light—First Beer in Space," YouTube video, 2:24, https://www.youtube.com/watch?v=_00eZtsuJ9M.

68. Foster Films, "ARDBEG—The Eyes of The World," Vimeo video, :07, https://vimeo.com/53365349.

69. Ben Foster (founder, Foster Film Company), in a conversation with the author, February 2019.

70. Foster Films, 0:31.

71. Foster Films, 1:12.

72. Foster Films, 1:17.

73. Peter Smith (planetary scientist), in an email conversation with the author, June 2018.

74. Robert Zubrin, The Case for Mars (New York: Free Press, 2011), 161.

75. Doug Messier, "Bob Zubrin, Mars & Beer," Parabolic Arc, last modified March 28, 2016, http://www.parabolicarc.com/.

76. Messier, "Bob Zubrin."

77. "NASA Scans Vineyards from Above to Help Growers." NASA press release, August 28, 2001, https://www.nasa.gov/.

78. NASA press release.

79. NASA Spinoff, "Professional Development Program Gets Bird's-Eye View of Wineries," https://spinoff.nasa.gov/Spinoff2017/cg_6.html.

80. NASA Spinoff article.

81. "ISS Benefits For Humanity: From NASA to Napa," NASA YouTube video, 3:44, https://www.youtube.com/.

82. NASA YouTube video, 4:06.

83. Kellie Gerardi (crew member, MDRS Crew #149), in an interview with the author, June 2018.

84. Richard J. Phillips (president, Phillips & Company), in a conversation with the author, July 2018.

85. Phillips interview.

86. Phillips interview.

87. Teressa Iezzi, "Red Bull CEO Dietrich Mateschitz On Brand As Media Company," Fast Company, last modified February 17, 2012, https://www.fastcompany.com/.

88. Iezzi, "Red Bull CEO."

89. Gerardi interview.

90. Nye interview.

## Chapter 6

1. Weir, The Martian, p. 21.

2. Raymond Wheeler (plant physiologist, NASA), in a conversation with the author, June 2018.

3. Jennifer Wadsworth, Charles S. Cockell, "Perchlorates on Mars Enhance the Bacteriocidal Effects of UV Light," Nature: Scientific Reports vol. 7, no. 4662 (July 2017), https://www.nature.com/articles/s41598-017-04910-3#Sec6.

4. Wheeler interview.

5. Ally Koehler, "Meet the Team Trying to Cultivate Spuds in Space," Red Bull, last modified May 22, 2017, https://www.redbull.com/

6. David A. Ramirez, Jan Kreuze, Walter Amoros, et al., "Extreme Salinity Is a Challenge to Grow Potatoes Under Mars-Like Soil Conditions: Targeting Promising Genotypes," International Journal of Astrobiology (2017), 1–7, https://www.cambridge.org/core/journals/international-journal-of-astrobiology/.

7. John Connolly (owner, Star Creek Vineyards; aerospace professional), in a conversation with the author, September 2018.

8. Wieger Wamelink (Wageningen University), in a conversation with the author, July 2018.

9. Koehler, "Meet the Team Trying to Cultivate."

10. Ramirez, "Extreme Salinity."

11. Chris McKay (planetary scientist, NASA), in an email conversation with the author, June 2017.

12. McKay interview.

13. Ramirez, "Extreme Salinity."

14. Associated Press in Lima, Peru, "'Super Potato' Grown in Mars-Like Conditions May Benefit Earth's Arid Areas," *The Guardian*, last modified March 30, 2017, https://www.theguardian.com/.

15. Franklin Briceno, "Lab Creates 'Super Potato' That Could Grow on Mars," *Orlando Sentinel*, last modified March 30, 2017, https://www.orlandosentinel.com/.

16. Linda Herridge, "NASA Plant Researchers Explore Question of Deep-Space Food Crops," NASA, last modified February 17, 2016, https://www.nasa.gov/.

17. Wamelink interview.

18. Wamelink interview.

19. Wamelink interview.

20. Wamelink interview.

21. Wamelink interview.

22. Wamelink interview.

23. Scripps Howard News Service, "Rocks Turned into 'Moon Dirt,'" *Chicago Tribune*, November 30, 1988, https://www.chicagotribune.com/.

24. Penelope Boston, David Blackmore, Reed Walker, Chris Carberry, "Libations in Space," Panel discussion, Explore Mars, Inc., Washington, D.C., May 11, 2017.

25. Gioia Massa, Raymond Wheeler, Robert Morrow, "Veggie Hardware Validation Test (Veg-01)," https://www.nasa.gov/.

26. "Veggie Hardware Validation Test (Veg-01)."

27. Scott Kelly (astronaut, NASA), in an email discussion with the author, June 2018.

28. Julia Sable, "Does Lettuce Taste Different in Space?," *Space Station Explorers*, last modified October 3, 2016, https://www.spacestationexplorers.org/.

29. "Space Station Live: Everything's Coming Up Veggie," NASA YouTube video, 2:46, April 13, 2016, https://www.youtube.com/watch?v=9JDAZBoLJUc.

30. Linda Herridge, "Veggie Plant Growth System Activated on International Space Station," NASA, last modified May 16, 2014, https://www.nasa.gov/.

31. NASA YouTube video.

32. Space Dynamics Laboratory, "Lada: Gardening in Space," https://www.sdl.usu.edu/downloads/lada.pdf.

33. Lori Meggs, "Growing Plants and Vegetables in a Space Garden," NASA, last modified June 15, 2010, https://www.nasa.gov/.

34. Meggs, "Growing Plants and Vegetables."

35. Christophe Lasseur (MELiSSA Project), in a conversation with the author, June 2018.

36. "Closed Loop Concept," ESA MELiSSA Project, last modified May 26, 2015, https://www.esa.int/.

37. "Planting Oxygen," ESA, last modified December 15, 2017, http://www.esa.int/.

38. "ESA's Melissa Life-Support Programme Wins Academic Recognition, ESA, last modified July 24, 2017, http://www.esa.int/.

39. "Water Recycling for Monks and Astronauts Awarded Dutch Innovation Prize," ESA, last modified December 21, 2018, http://www.esa.int/.

40. ESA, "Water Recycling for Monks."

41. Lasseur interview.

42. Lasseur interview.

43. "Life Support Pilot Plant Paves the Way to Moon and Beyond," ESA, last modified June 5, 2009, "https://www.esa.int/.

44. Pascal Lee (Mars Institute), in an email conversation with the author, December 2018.

45. "Arthur Clarke Mars Greenhouse (ACMG): Frequently Asked Questions," SpaceRef press release, July 18, 2002, http://www.spaceref.com/.

46. Pascal Lee interview.

47. Lee interview.

48. Raymond Wheeler, "Agriculture for Space: People and Places Paving the Way," *Open Agriculture* vol. 2 (2017): 14–32.

49. Taber MacCallum (former crew member, Biosphere 2), in a conversation with the author, June 2018.

50. MacCallum interview.

51. "VP, Abu Dhabi Crown Prince launch Mars Science City," UAE press release, September 26, 2017, http://mediaoffice.ae/en/mediacenter/.

52. UAE press release.

53. Russell Hotten, "Dubai Airshow: Why the UAE Is Probing Space Agriculture," *BBC News*, last modified November 15, 2017, https://www.bbc.com/.

54. Hotten, "Dubai Airshow."

55. Leonard David, "China Wraps Up 1-Year Mock Moon Mission to Lunar Palace 1," Spacewww, last modified May 17, 2018, https://www.space.com/.

56. *GB Times*, "Lunar Palace-1: A Look Inside China's Self-Contained Moon Training Habitat," *GB Times*, last modified May 16, 2018, https://gbtimes.com/.

57. Jamie Ayque, "Chinese Astronauts Are Growing Rice on the Tiangong Space Station," *Nature World News*, last modified November 18, 2016, https://www.natureworldnews.com/.

58. Yasmin Tayag, "China Is About to Land Living Eggs on the Far Side of the Moon," *Inverse*, last modified January 2, 2019, https://www.inverse.com/.

59. "Mars Society Launches Translife Mission Project," Mars Society press release, August 30, 2001, http://www.spaceref.com/.

60. "Eu:CROPIS—Greenhouses for the Moon and Mars," DLR press release, May 24, 2016, https://www.dlr.de/dlr/en/.

61. Penelope Boston, David Blackmore, Reed Walker, Chris Carberry, "Libations in Space," Panel discussion, Explore Mars, Inc., Washington, D.C., May 11, 2017.

62. Connolly interview.

63. "Libations in Space."

64. Blackmore interview.

65. Blackmore interview.

66. Wheeler interview.

67. "A Conversation with E.O. Wilson," *PBS NOVA*, April 1, 2018, https://www.pbs.org/.

68. Wheeler interview.

69. Wheeler interview.

70. Nye interview.

71. Lynne Rothschild, "Synthetic Astrobiology: Synthetic Biology Meets ET," slide presentation, http://www.ucmp.berkeley.edu/about/shortcourses/Rothschild%20Synbio%20meets%20ET%20Newcastle%20UCMP%20022018_small.pdf.

72. Lynn Rothschild, "Synthetic Biology Meets Bioprinting: Enabling Technologies for Humans on Mars (and Earth)," *Biochemical Society Transactions* vol. 44, no. 4 (2016): 1158–1164.

73. Rothschild, "Synthetic Biology Meets Bioprinting."

74. Matt Simon, "Lab-Grown Meat Is Coming, Whether You Like It or Not," *Wired*, last modified February 16, 2018, https://www.wired.com/.

75. Simon, "Lab-Grown Meat Is Coming."

76. "Lynn Rothschild Talks About Creating Synthetic Organisms and NASA's Search for Life Beyond Earth," NASA, December 21, 2016, https://www.nasa.gov/.

77. Simon "Pete" Worden (former Director, NASA Ames Research Center), in an email conversation with the author, January 2018.

78. "Libations in Space."

79. Jessica Kozuka, "Lettuce Makes Tasty Booze and Other Discoveries at a New All-You-Can-Drink Shochu Bar," *SoraNews24*, last modified July 7, 2016, https://soranews24.com/.

80. Preston Phro, "We Got Some Japanese Radish Sparkling Wine, But Didn't Expect It to Taste Like This," *SoraNews24*, last modified November 5, 2013, https://soranews24.com/.

81. National University of Singapore, "World's First Alcoholic Beverage Made from Tofu Whey," *Science Daily*, last modified November 27, 2017, https://www.sciencedaily.com/.

82. Regan Stephens, "Finally, Carrots Are Available in Alcoholic Beverage Form—Just in Time for Easter," *Food & Wine*, last modified March 30, 2017, https://www.foodandwine.com/.

83. Spencer interview.

84. Worden email interview.

85. Dixon interview.

## Chapter 7

1. Neil deGrasse Tyson (astrophysicist), in an email conversation with the author, February 2019.

2. "Cosmic Cloud Contains Enough Alcohol to Keep the Whole World Drinking for a Billion Years," *ScienceAlert*, July 25, 2014, https://www.sciencealert.com/.

3. David Spiegelhalter, "The Risks of Alcohol (Again)," University of Cambridge, August 24, 2018, https://medium.com/wintoncentre/the-risks-of-alcohol-again-2ae8cb006a4a.

4. Kluger interview.

5. Anderson interview.

6. Forsyth, 171.

7. Forsyth, 171.

8. Saralyn Mark interview, January 2019.

9. Nye interview.

10. Nye interview.

11. Forsyth, 230–231.

# Bibliography

Adams, Douglas. *The Hitchhiker's Guide to the Galaxy*. New York: Ballantine Books, 2005. Kindle edition.

Adams, Douglas. *The Restaurant at the End of the Universe*. New York: Ballantine Books, 2005.

"Alcosynth: The Hangover-Free Future of Drinking." YouTube. Video, 1:34. December 13, 2017. https://www.youtube.com/watch?v=J0-JEh5mZ7E.

Aldrin, Buzz, and Ken Abraham. *Magnificent Desolation: The Long Journey Home from the Moon*. New York: Harmony Books, 2009.

American Home Brewers Association. "The History of Beer at Oktoberfest," https://www.homebrewersassociation.org/.

Anson, Jane. "Anson on Thursday: The Surprising Adventure of Lynch Bages in Space." *Decanter*, June 18, 2015. https://www.decanter.com/wine-news/.

Associated Press in Lima, Peru. "'Super Potato' Grown in Mars-Like Conditions May Benefit Earth's Arid Areas." *The Guardian*, March 30, 2017. https://www.theguardian.com/.

Baum-Baiker, Cyhthia. "The Health Benefits of Moderate Alcohol Consumption: A Review of the Literature." *Drug and Alcohol Dependence* 15, no. 3 (1985): 207–227.

Benford, Gregory. "All the Beer on Mars." In *Isaac Asimov's Scientific Fiction* 13, no. 1 (January 1989): 61.

Benford, Gregory. "A Fairhope Alien." In *Stories of the Blue Moon Café II*, edited by Sonny Brewer. San Francisco: MacAdam/Cage, 2003.

"Benjamin Rush's Rules on Health." From the *Letters of the Lewis and Clark Expedition with Related Documents*, 1978. https://www.lewisclark.org/article/2394.

Borman, Frank. "Oral History, Interview with Catherine Harwood, April 13, 1999." In Las Cruces, New Mexico, Transcript, NASA Johnson Spaceflight Center Oral History Program. https://www.jsc.nasa.gov/history/oral_histories/.

Bourland, Charles T. Oral History. Johnson Space Center Oral History Project, April 2006. www.jsc.nasa.gov/history.

Bourland, Charles T., and Gregory L. Vogt. *The Astronaut's Cookbook: Tales, Recipes, and More*. New York: Springer, 2010.

Bradbury, Ray. *The Martian Chronicles*. New York: Simon & Schuster, 1977.

Brin, David. "Talk Farm Dynamo." 1983. http://www.davidbrin.com/tankfarm.htm.

Bunch, Joey. "Rep. Soggy Sweat could guide Colorado lawmakers on whiskey." *The Spot, The Denver Post*, April 26, 2016. http://blogs.denverpost.com/thespot/.

Burgess, Colin. *Sigma 7: The Six Mercury Orbits of Walter M. Schirra, Jr*. Chichester, UK: Praxis Publishing, 2016.

Bustard, Bruce. *Spirited Republic: Alcohol's Evolving Role in U.S. History*. 2014. https://www.archives.gov/.

Butov, Mikhail. "When the Tsar Banned Booze." *Russia Today*, August 18, 2014. https://www.rbth.com/.

Clark, Joe, and Stuart Derrick. *The Ultimate Book of Whiskey*. New York: Paragon Books Ltd., 2014.

Cormier, Zoe. "The End of Hangovers." Medium.com, December 21, 2017. https://medium.com/neodotlife/.

Damerow, Peter. "The Origins of Brewing Technology in Ancient Mesopotamia." *Cuneiform Digital Library Journal* (January 22, 2012). https://cdli.ucla.edu/.

Davis, Marni. *Jews and Booze: Becoming American in the Age of Prohibition*. New York: NYU Press, 2012.

Debczac, Michele. "The Time Coke and Pepsi Brought Their Rivalry to Outer Space." *Mental Floss*, September 10, 2015. http://mentalfloss.com/.

Dunbar, Robin. "Why Drink Is the Secret to Humanity's Success." *Financial Times,* last modified August 10, 2018. https://www.ft.com/.

Dunn, Marcia. "Fire, Collision, Noxious Fumes, and Cognac—All in a Day's Work on Mir." *L.A. Times,* February 22, 1998. http://articles.latimes.com/.

European Space Agency (ESA). "50 Years of Humans in Space: Gagarin's Traditions." https://www.esa.int/.

Ferris-Rotman, Amie. "White Wine on the Red Planet? Scientists in Georgia Are Hunting for a Perfect Martian Grape." *The Washington Post,* January 7, 2019. https://www.washingtonpost.com/.

Foale, Michael. "What It's Like to … Spend a Year Floating in Space." *The Telegraph,* October 23, 2013. https://www.telegraph.co.uk/.

Forster, Katie. "'Hangover-Free Alcohol' Could Replace All Regular Alcohol by 2050, Says David Nutt." *Independent,* September 22, 2016. https://www.independent.co.uk/.

Forsyth, Mark. *A Short History of Drunkenness.* New York: Three Rivers Press, 2017.

Franklin, Benjamin. *A Benjamin Franklin Reader,* edited and annotated by Walter Isaacson. New York: Simon & Schuster Paperbacks, 2003.

Gately, Iain. *Drink: A Cultural History of Alcohol.* New York: Gotham Books, 2008.

Hafiz, Yasmine. "The Moon Communion of Buzz Aldrin That NASA Didn't Want to Broadcast." *Huffington Post,* December 6, 2017. https://www.huffingtonpost.com/.

Hames, Gina. *Alcohol in World History.* London: Routledge, 2014.

Harrell, Jason. *The Bartending Therapist.* Amazon Digital Services, 2016.

Higgins, E. Arnold, Audie W. Davis, John A. Vaughan, et al. "The Effects of Alcohol at Three Simulated Aircraft Cabin Conditions." Federal Aviation Administration, September 1968.

Higgins, E. Arnold, John A. Vaughan, Gordon E. Funkhouser, et al. "Blood Alcohol Concentrations as Affected by Combinations of Alcoholic Beverage Dosages and Attitudes." Federal Aviation Administration, April 1970.

Hill, Arthur. "Astronauts Finally Get Out of this World Wine." *The Milwaukee Journal,* August 1, 1972.

Hilt, David, Owen Garriott, Joe Kerwin. *Homesteading Space: The Skylab Story.* Lincoln: University of Nebraska Press, 2008.

Hornsey, Ian S. *Alcohol and Evolution of Human Society.* Cambridge, U.K.: RSC Publishing, 2012.

Hotten, Russell. "Dubai Airshow: Why the UAE is Probing Space Agriculture." *BBC News,* November 15, 2017. https://www.bbc.com/.

"How Do You Make a Whisky Glass for Space?" Medium.com, September 2, 2015. https://medium.com/space-glass/.

Huntford, Roland. *Shackleton.* New York: Carroll & Graf, 1998.

Jefferson, Thomas. *The Papers of Thomas Jefferson: Retirement Series,* edited by J. Jefferson Looney. Princeton: Princeton University Press, 2016.

Johnson, Hugh. *The Story of Wine.* London: Mitchell Beazley, 1989.

Jones, Graham. "Mumm Grand Cordon Stellar—Champagne Designed for Space?" https://www.apetogentleman.com/mumm-grand-cordon-stellar/.

Jude, Deborah L. "Can You Brew Beer on the Moon?" University of California, San Diego, January 19, 2017. https://www.universityofcalifornia.edu/news/.

Kelly, Scott. *Endurance.* New York: Knopf, 2017.

Koehler, Ally. "Meet the Team Trying to Cultivate Spuds in Space." *Red Bull,* May 22, 2017. https://www.redbull.com/.

Lacovara, Peter. *The World of Ancient Egypt: A Daily Life Encyclopedia.* Santa Barbara: Greenwood, 2017.

Lane, Edward William. *An Account of the Manners and Customs of Modern Egyptians.* London: John Murray, Albemarle Street, 1860.

Leonard, David. "Space Farming: Satellite's Greenhouses to Simulate Moon, Mars Gravity." Spacewww, February 1, 2017. https://www.space.com/.

Levkovich, Yevgeny. "Space Smugglers: How Russian Cosmonauts Sneak Booze into Outer Space." *Russian Beyond,* April 12, 2017. https://www.rbth.com/.

Lufkin, Bryan. "Why Astronauts Are Banned from Getting Drunk in Space." *BBC,* February 2017. http://www.bbc.com/.

Lumsden, Bill. "The Impact of Microgravity on the Release of Oak Extractives into Spirit." Ardbeg white paper, 2015. https://www.ardbeg.com/sites/ardbeg.com/.

Malone, Noreen. "Undercover at the U.N. Lounge, Where Diplomats Get Drunk and Handsy." *The New Republic,* January 30, 2014. https://newrepublic.com/.

Martin, Alan. "Alcohol in Space: From Communion Wine to Zero-Gravity Whisky." ALPHR, September 5, 2015. https://www.alphr.com/.

Mather, Increase. *Wo to Drunkards.* Boston: Tomothy Green, at the Lower End Middle-Street, 1712.

McGovern, Patrick E. et al. "Fermented Beverages of Pre- and Proto-Historic China." *Proceedings of the National Academy of Sciences,* December 2004. https://www.pnas.org/.

Meggs, Lori. "Growing Plants and Vegetables in a Space Garden." NASA, June 15, 2010. https://www.nasa.gov/.

Messier, Doug. "Bob Zubrin, Mars & Beer." *Parabolic Arc,* March 28, 2016. http://www.parabolicarc.com/.

Murcia, Francisco Javier. "Wine, Women, and Wisdom: The Symposia of Ancient Greece." *National Geographic* (January/February 2017).

NASA. "NASA Astronaut Health Care System Review Committee, February—June, 2007, Report to the Administrator." https://www.nasa.gov/.

NASA. "PART III: Skylab Development and Operations: February 1970-November 1974." *SP-4011 Skylab: A Chronology,* August 10, 1972. https://history.nasa.gov/.

NASA Spinoff. "Professional Development Program Gets Bird's-Eye View of Wineries." https://spinoff.nasa.gov/Spinoff2017/cg_6.html.

NASA. "Suds in Space." September 21, 2001. https://science.nasa.gov/.

O'Connor, Bryan (NASA Chief Safety & Mission Assurance Officer). "NASA Safety Review." *C-Span,* August 29, 2007. https://www.c-span.org/.

Peele, Stanton, and Archie Brodsky. "Exploring Psychological Benefits Associated with Moderate Alcohol Use." *Drug and Alcohol Dependence,* 2000.

Plack, Noelle. "Intoxication and the French Revolution." *Age of Revolution,* December 5, 2016. https://ageofrevolutions.com/.

Ramirez, David A., Jan Kreuze, Walter Amoros, et al. "Extreme Salinity is a challenge to Grow Potatoes Under Mars-Like Soil Conditions: Targeting Promising Genotypes." *International Journal of Astrobiology* (2017): 1–7. https://www.cambridge.org/core/journals/international-journal-of-astrobiology/.

Roberts, Sam. "How the Jews Handled Prohibition." *New York Times,* January 27, 2012.

Robinson, Spider. *The Callahan Chronicles.* New York: A Tom Doherty Associates Book, 1997.

Romanoff, Jim. "When It Comes to Living in Space, It's a Matter of Taste." *Scientific American,* March 10, 2009. https://www.scientificamerican.com/.

Rothschild, Lynn. "Synthetic Biology Meets Bioprinting: Enabling Technologies for Humans on Mars (and Earth)." *Biochemical Society Transactions* 44, no. 4 (2016): 1158–1164.

Rush, Benjamin. *An Inquiry into the Effects of Ardent Spirits Upon the Human Body and Mind.* New York: Printed for Cornelius Davis, 1811.

Sable, Julia. "Does Lettuce Taste Different in Space?" Space Station Explorers, October 3, 2016. https://www.spacestationexplorers.org/.

Sagimoto, Manabu, Elena Shagimardonova, Oleg Gusev, et al. "Gene Expression of Barley Grown in Space." *Space Utiliz Res* 24 (2008). https://repository.exst.jaxa.jp/

Scripps Howard News Service. "Rocks Turned into 'Moon Dirt.'" *Chicago Tribune,* November 30, 1988. https://www.chicagotribune.com/.

Seal, Laura. "Mumm to Launch Space Champagne for Astronauts." *Decanter,* last modified June 8, 2018. https://www.decanter.com/.

"Shipwreck Champagne Sells for $156,000, Preserved For 170 Years Underwater." *Huffington Post,* June 8, 2012. https://www.huffingtonpost.com/.

Simon, Matt. "Lab-Grown Meat Is Coming, Whether You Like It or Not." *Wired,* February 16, 2018. https://www.wired.com/.

Spotts, Peter. "Worms in Space: How One Experiment Could Send Them to Mars." *Christian Science Monitor,* November 30, 2011. https://www.csmonitor.com/.

Sterrett, Kirsten, Marvin Luttges, Steven J. Simske, et al., "A Study of Saccharomyces Uvarum Fermentation in a Microgravity Environment." *Technical Quarterly* 3, no. 1 (1996): 33–38.

Taylor, Kate. "Millennials are Creating a Mounting Crisis for Some of the Most Iconic Beer Brands in America." *Business Insider,* August 4, 2018. https://www.businessinsider.com/.

Thagard, Norman. Interview with Rebecca Wright, Paul Rollins, Carol Butler, September 16, 1998. Shuttle-Mir Oral History Program. https://spaceflight.nasa.gov/history/.

Torbati, Yeganeh. "U.S. and French Navies Differ on Alcohol Aboard Ships." Originally by *Reuters,* December 19, 2015. https://gcaptain.com/.

Twilley, Nicola Twilley. "Why Astronauts Were Banned from Drinking Wine in Outer Space." *GIZMODO,* January 31, 2014. https://gizmodo.com/.

UAE. "VP, Abu Dhabi Crown Prince Launch Mars Science City." UAE press release, September 26, 2017. http://mediaoffice.ae/en/media-center/.

Verne, Jules. *From Earth to the Moon.* New York: Charles Scribner's Sons, 1890.

Wadsworth, Jennifer, and Charles S. Cockell. "Perchlorates on Mars Enhance the Bacteriocidal Effects of UV Light." *Nature: Scientific Reports* 7, no. 4662 (July 2017). https://www.nature.com/articles/s41598-017-04910-3#Sec6.

Weir, Andy. *Artemis*. New York: Crown, 2017.

Wells, H.G. *The First Men in the Moon*. Indianapolis: The Bowen-Merrill Company, 1901.

Wheeler, Raymond. "Agriculture for Space: People and Places Paving the Way." *Open Agriculture* 2 (2017): 14–32.

Wii, Sylvia. "China Grows Wine in Space to Beat Harsh Climate." *Decanter*, September 20, 2016. https://www.decanter.com/.

Yasutaka, Daimon. "Sake—Drink of the Gods, Drink for the People." Lecture at Japan Society, New York, NY, October 12, 2000. http://esake.com/Brewers/DaimonB/Naorai/naorai.html.

Zubrin, Robert. *The Case for Mars*. New York: Free Press, 2011.

# Index

Numbers in **bold italics** indicate pages with illustrations

Hyde de Neuville, Jean Guillaume 32
Hyslop, Sandy 137

Ichikawa, Junichi 98
International Potato Center 151
International Space Station 8, 10, 70–71, 80–
  82, 92, 97–98, *100*–101, 114, 120–125, 131–
  *132*, 139, 147, 151, 158–160, *162*–163, 175
Inuits 14
Ireland 22
Islam 24–25

Japan 25–26, 96–98, 123–125; *see also* Sap-
  poro; Shintoism; Suntory
Jefferson, Thomas 30–32
Joe's Bar 57–*59*; *see also Battlestar Galactica*
Johnson, Hugh 19
Jose Cuervo *140*–141
J.P Aerospace 141
Judaism 23–24, 36

Kaylan, Srivaths 114
Kelly, Scott *83*, 85, *159*–160; *Endurance* 84–85
Kerwin, Joe: *Homesteading Space: The Skylab
  Story* 75
King James IV of Scotland 22
Kloeris, Vickie 92, 101
Kluger, Jeffrey 90, *132*, 185
Kreuze, Jan 151

Lada 98, 160–161; *see also* International Space
  Station
Lane, Edward William 24
Lasseur, Christophe 161, 163–165
Lazutkin, Alexander 75
Leclercq, Gerard 122
Lee, Pascal 89, 165–166
Leonov, Alexey 84
Lewis, Merriweather 33–34
Limdgren, Kjell *159*–160
Linenger, Jerry 80
Liu, Hong 169
Lovell, James A., Jr. *65*–*66*
Lumsden, Bill 120–122

MacCallum, Taber 167–168
Maenads 17–18
Maison Mumm 125–128, 133
Mallya, Vijay 35
Mamedov, Marat 181
Manber, Jeffrey 75–77, 81, 119–123
Mara, Kate *99*
Mariotti, Dider 126
Mark, Saralyn 70–71, 188
Marques, Ricardo 98–*99*, 102–103
Mars Desert Research Station (MDRS) 115–
  118
Mars Institute 165
Mars Science City 168–169
Mars 2117 Project 168
Mars World Enterprises 181–182

Marsarita *140*–141
Massa, Gioia 160
Mateschitz, Dietrich 146–147
Mather, Increase: "Wo to Drunkards" 30
*The Mayflower* 29
Mazzei, Philip 32
McElveen, Ryan 49–50
McKay, Chris *88*–89, 153
MELiSSA 161–165; *see also* International
  Space Station
Mercury Program 64
Milliways 58; *see also The Restaurant at the
  End of the Universe*
Mir space station 8, 71, 75–82, 84, 91, 160; fire
  and collision 80; *see also* Russia
Mitchell, Jaron 104
mold reduction 144–145
Mondavi, Tim 142
Moon: beer 113–115; regolith *154*–158; wine
  making (sci fi) 45
Morellet, Abbé 33
Mos Eisley cantina 50–*51*
Mullane, Mike *83*
Musk, Elon 10, 50, 102, 115, *171*

Nanoracks 120–123
Napa, California 144–145
NASA 8, 10, 13, 43, 45, 62–*65*, *67*–68, 71–74,
  76–79, 81–*83*, 85–86, 91–93, 102, 105, 109,
  *112*, 115, 120, 131, 133, 142–144, 150, 152–153,
  157–160, 165. 176–178, 188; The Breach
  House 82–*83*
NASA Ames Research Center 10, 152
NASA Astronaut Health Core System Review
  Committee 85–87
Natural Light 138–139
Nerima Daikon 179
The IX Millennium Project 129–131
Ninkasi Brewing Company 109–113
Norse mythology 20
Nowak, Lisa 86–87
Nutt, David 137–138
Nye, Bill 129, 133, 147–148, 176, 189

O'Connor, Bryan 86
Okayama University 97
Olson, Greg 131–*132*
O'Neill, Gerald K. *171*
O'Neill, Ian 41, 57, 102
Open Space Agency 135–137
Orion crew capsule 10

Panarin, Igor 86
Pancho's Happy Bottom Riding Club 82
Parazynski, Scott 92
Parr, James 135–137
*Passengers* 54–55
Paul Masson California Rare Cream Sherry
  72–*73*
Peele, Stanton 27
Pepsi 91

Index 209